PRACTICAL GUIDE TO INTERNATIONAL STANDARDIZATION FOR ELECTRICAL ENGINEERS

PRACTICAL GUIDE TO INTERNATIONAL STANDARDIZATION FOR ELECTRICAL ENGINEERS

IMPACT ON SMART GRID AND E-MOBILITY MARKETS

Hermann J. Koch
Gerhardshofen, Germany

WILEY

This edition first published 2016
© 2016 John Wiley & Sons, Ltd

Registered Office
John Wiley & Sons, Ltd, The Atrium, Southern Gate, Chichester, West Sussex, PO19 8SQ, United Kingdom

For details of our global editorial offices, for customer services and for information about how to apply for permission to reuse the copyright material in this book please see our website at www.wiley.com.

Library of Congress Cataloging-in-Publication data applied for

ISBN: 9781119067412

A catalogue record for this book is available from the British Library.

Set in 10/12pt Times by SPi Global, Pondicherry, India

Printed and bound in Malaysia by Vivar Printing Sdn Bhd

1   2016

Contents

Foreword by Mark Waldron

At first glance, standards, and particularly international standards, appear to have a very clear and singular function: to define widely applicable technical performance requirements within their scope of application. Of course this **is** a key purpose of standards but the role and influence of standards and the standardization process are much broader than they might at first appear. Knowledge of, and engagement with, standards and the processes by which they are produced is invaluable for engineers and scientists throughout the life cycle of any product of service, from research and development, through production and service, to end of life and disposal.

Development of new standards should always take place to address a market need. This need may derive directly from a customer desire to implement new technology, services or facilities; it may be driven by developments in technology within the suppliers of products and services but more typically it is a combination of these factors. In any case, prestandardization activities will typically be undertaken among international experts to establish the maturity of the intended field of standardization and to establish a common language, definitions and mutual understanding within the field. Whether done internally by standardization bodies or externally by organizations such as CIGRE this process of establishing a common language and understanding is key to effective standardization. Prestandardization activity also commonly highlights areas where critical knowledge is lacking and thereby provides feedback for further research and development required prior to the establishment of a standard or standards. Finally prestandardization can also identify aspects that should not be Standardized, for example because there is no common approach possible or because they are subject to specific local requirements.

Once initiated, a key strength of the standardization process itself is that it brings together a wide range of stakeholders with a need to establish clear, unambiguous requirements that are deliverable (at reasonable cost) and are mutually acceptable to all. Since standards address aspects such as technical performance, operation and operational facilities, safety, environmental impact, testing and interoperability, it is common for researchers, designers, manufacturers, testing facilities, users, regulators and consultants to be engaged in their development. As well as resulting in an effective standard, this process provides every participant with a valuable insight into the perspectives of other stakeholders in the field, which is difficult to gain effectively by other means. It is also a great training ground in the arts of negotiation and compromise!

Finally, even the best standard will have scope for improvement once it has been applied and used by a wide range of stakeholders. Feedback into the standard-making process from the widest possible stakeholder base is vital to ensure the best possible standard and to ensure that developments within the scope of application are addressed.

So, in summary, standards and the processes by which they are prepared have a considerable influence on the activities of engineers working in the field of electrical engineering and a knowledge of, and ideally participation in, this activity is undoubtedly advantageous and may even be considered essential.

Mark Waldron
CIGRE TC Chairman

Foreword by Bernhard Thies

Modern societies would not work properly without standards. From basic commodities like bulbs or a sheet of paper to highly complex machineries and power plants: Nothing runs without technical rules. Norms and standards as commonly recognized state of the art lay down not only interfaces as precondition for exchangeability, comparability and interoperability. The user independent of being a consumer or an integrator also obtains assurance regarding the required level of safety and quality.

In this manner the term safety means to comprehensively protect humans, animals and objects against any harm regardless of the threat scenario. The key is to already take the necessary precautions at the design phase of a new product to reduce any risk to a minimum. A standard represents the extensive experience of many experts – engineers, scientists, safety experts, environmentalists as well as consumer advocates. If a product fulfills the requirements of such a consensus-driven standard a high level of safety is automatically classified. Designers and developers benefit from the standard in the way that their work becomes more efficient and reliable. Thereby, the standard only provides basic requirements so that there is still enough space for innovation and creativity. Hence, standards by no means impede innovation but lay down a level playing field on which competitors can build different solutions with unique selling propositions.

However, standardization requires the input of many experts that provide their knowledge for the common property. Moreover, companies delegating experts into standardization bear the costs of travelling and personnel. But, companies also benefit from direct participation within standard committees by shaping the standard to their advantage or gaining knowledge prior to the publication of a standard. To sum up, not only the individual standard setters, but also the whole society benefits from standardization since standards promote technological acceptance and open-mindedness. Standardization can achieve a highly operational and economical benefit which is estimated around 16 billion Euros per year for Germany.

Bernhard Thies
Chairman of the Board of Directors
DKE Technical Standardization
Electric, Electronic and Information Technology
Frankfurt, Germany

Foreword by Markus Reigl

Many assertions are made about standardization and standards – and the most of them are true! Now let us take a look at them from various perspectives.

Firstly, from a governmental perspective, standards support regulatory requirements and help to achieve societal goals such as safety in operation, user and environmental friendliness, energy efficiency and sustainability. Further, standards set the scene by stipulating the commonly accepted basic requirements that various vendors have agreed on. These same vendors compete in markets based on product features, performance, quality and price. Through this mechanism standards help to intensify competition.

If true international standards are widely adopted in global target markets the major advantage for vendors using the standards is to capitalize on their broad market acceptance so reducing country specific re-design or re-engineering.

Finally product users benefit from the extensive variety of products made by different vendors and at the same time they can be confident with the conformity to legal regulations. In addition they benefit from interoperability in heterogeneous multi-vendor solutions. Furthermore these standards provide investment security from simple machinery to complex large scale industrial plants.

After extolling all the merits of using standards we should not however forget to honor those who make them – the innumerable technical experts in the committees and working groups of standards developing organizations. Any such committee can consider itself more than fortunate if it has highly skilled, knowledgeable and experienced industry experts contributing to its standardization work. Experts such as Dr. Hermann J. Koch.

I can thoroughly recommend Hermann J. Koch's practical guide which provides "hands on" expert knowledge. The international standardization community would benefit greatly if there were more key experts like Hermann J. Koch. Enjoy the guide.

Markus Reigl, Dipl-Ing, MA
Head of the Corporate Department for
Technical Regulation and Standardization
Siemens AG
Berlin and Munich, Germany

Foreword by Damir Novosel

Major technological innovations in the areas such as renewable energy resources, storage, electric vehicles, automation, measurement devices, protection and control, materials, DC technology and robotics resulted in a paradigm shift of how we use electricity. The electric power and energy industry is in a crucial transition phase as initiatives we take today will affect how the grid is operated for years to come. In this fast-pace environment, standards are even more critical for both users and vendors to streamline deployment of both existing and new technologies and support interoperability among devices and systems as well as the use of best industry practices.

Active participation in development of Standards has been helping our membership to enhance and protect current and future investments, shape industry practices, and influence new developments. IEEE members need to be even more engaged and with support and leadership from the IEEE Standards Association continue working diligently to better serve our industry in releasing standards in timely fashion.

As we emphasize importance of IEEE standards and technical reports, it is important to remember that they have been providing fundamental value to our industry since the dawn of electricity. Figure below shows first AIEE (IEEE predecessor) standard published in 1893.

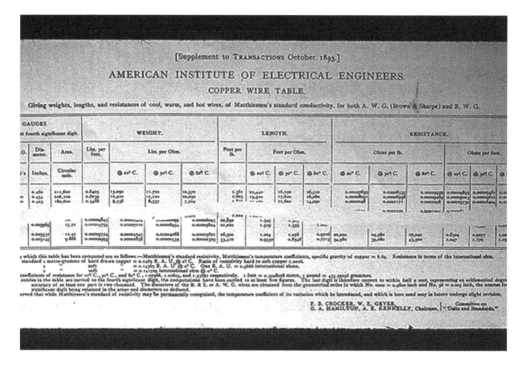

Figure 1   First IEEE (AIEE) Standard

Presently, a lot of countries in the world have industry regulations/codes based on IEEE related standards. The goal of IEEE, including IEEE Power and Energy Society (PES) which publishes over 40% of IEEE standards, is to continue developing required standards and focus on promoting them globally.

This book by one of the industry leaders in developing standards, Dr. Herman Koch, is very important to raise the awareness and communicate importance of standards, including recent developments.

Damir Novosel
IEEE PES President

Preface

Standardization today is a complex business. With influences at international, regional and national level it is like an ever-moving target and it is hard to follow, with its own processes. At the same time standardization is becoming increasingly important for the management of successful technical innovation and new products and services. In modern business strategies, Having the right standards in place when a new product or service is offered to the market is a key factor in the success of modern business strategies. Innovation may be the basis for success but the standard related to the innovation will open the market for the new product or service internationally, in a region, or in one country. As Werner von Siemens said in the late 1800s: 'He who owns the standards owns the market!' This is still valid today.

Standardization is changing fast and continuously adapting to market situations following market trends. One big goal of recent decades in international standardization was to reduce the time it takes to finish a standard. Standardization organizations developed new standard products like the publicly available specification (PAS). Another topic in recent years has been the trend towards globalization in industry. Standardization needs to keep up with this. Many national or regional organizations became international and opened new offices all over the world. The German DIN was used in Asia and South America; the British BSI was rolled out to all continents and the American-based IEEE opened offices in Europe, Asia, Africa and South America. All these activities influence the availability and acceptance of standards by users in the region. New types of standardization organizations based on industrial consortia create new standards in a fast-changing market of new technical products such as smart phones and software services. Most recently, the Internet of things has led to standards organized on the Internet, Linux software being an impressive example. All of this influences the impact of technical standards on new products and services.

The author is an active participant for more than 25 years for Siemens high voltage division. He is participating in standardization at the international, regional and national level in the field of electrical technology. Based on his experiences with IEC, IEEE, DKE and CIGRE and personal contact with other standardization organizations in France, Netherlands, the United Kingdom, Denmark, Russia, China, Japan, the United States, Canada, Brazil, South Africa,

Egypt, India and other countries, this book has been written to provide guidance and an overview of the subject. It also helps the reader to evaluate standardization activities.

The book gives a quick understanding of how standardization organizations work, how they are structured and how participation in standardization work is possible. It also provides useful information on general aspects of standardization.

Because of the nature of standardization, standardization activities and plans must be set up directly with the related standardization organization. The author cannot accept liability in relation to information given by this book.

Hermann J. Koch
Gerhardshofen

Acknowledgements

This book reflects my experience of international, regional and national standardization gained over 25 years. Contributions to this book came from many experts in the field.

I would first like to acknowledge my secretary Angela Dietrich, for her writing work, and Ulrich Ballas for creating all the graphics in the book. Without their support I would not have been able to finish the book on schedule.

Much of the material in the book has been taken from my university lecture at the Hochschule für Technik und Wirtschaft (HTW) Berlin – a compilation of information mainly provided by the Deutsches Institut für Normung (DIN), the Verband der Elektrotechnik, Elektronik und Informationstechnik (VDE) and the Deutsche Kommission Elektrotechnik Elektronik Informationstechnik im DIN und VDE (DKE) in Germany, the British Standards Institute (BSI), the American National Standards Institute (ANSI) and Institute of Electrical and Electronics Engineers (IEEE) in the United States, the Association Française de Normalisation (AFNOR) in France, the Nederlands Normalisatie Instituut (NEN) in the Netherlands, the Asociación Española de Normalización y Certificación (AENOR) in Spain, the Italian Comitato Elettrotecnico Italiano (CEI) and the Ente Nazionale Italiano di Unificazione (UNI) in Italy, the Russian Federal Agency on Technical Regulation and Metrology (GOST), the Standardization Association of the People's Republic of China (SAC), the Bureau of Indian Standards (BIS), the Japanese Standard Association (JAS) and the Standards Council of Canada (SCC).

The material has been collected using information from many Internet sites. Not everyone who has contributed to this book can be mentioned here but I would like to acknowledge the following in alphabetical order:

Luc Barranger, France AFNOR; Jean-Marc Biasse, France; Anne Bosma, Sweden; Wolfgang Brodt, Austria; Sivaji Chakravorti, India; Enrico Colombo, Italy; Terry Decourcelle, IEEE, USA; Denis Dufournet, France; Edgar Dullni, DKE, Germany; Jens Erdmann, Belgium; John Finn, United Kingdom; Kenneth Gettman, United States; Judith Gorman, IEEE, United States; Jodi Haasz, IEEE, United States; Tony Headley, United Kingdom; Guido Heit, DKE, Germany; Hisatoshi Ikeda, Japan; Gerhard Imgrund, DKE, Germany; Chris Jones, United Kingdom; Motofumi Matsumura, Japan; Enrique Otegui, Spain; Wan Ki Park, Korea; Patrick

Ryan, IEEE, United States; Gerard Schoonenberg, Netherlands; Bernhard Thies, DKE, Germany; Kyoichi Uehara, Japan and Willem Wolf, NEN Netherlands.

My colleagues in Siemens were Sven Achenbach, Heiko Englert, Peter Glaubitz, Matthias Gommel, Thomas Hammer, Friedrich Harless, Dirk Helbig, Claus Kern, Hartmut Knobloch, Edelhard Kynast, Peter Menke, Ansgar Müller, Markus Reigl, Heinz-Helmut Schramm, Ralph Sporer and Norbert Trapp.

Support from my family helped me to write the book and motivated me to bring it to a successful end. Thanks to my wife Edith, my son Christian and friend Britta, my daughter Katrin and friend Christopher for their support and the design of the front cover of the book.

Abbreviations

AA	DIN Arbeitsausschuss (committee)
AAL	Ambient Assisted Living
ABNT	Associação Brasileira de Normas Técnicas
AC	alternating current
AC	IEC Advisory Committee
AC ART	IEC Advisory Committee on Applications of Robot Technology
AC EA	IEC Advisory Committee on Environmental Aspects
AC EC	IEC Advisory Committee on Electromagnetic Compatibility
AC EE	IEC Advisory Committee on Energy Efficiency
AC OS	IEC Advisory Committee on Safety
AC SEC	IEC Advisory Committee on Security
AC TAD	IEC Advisory Committee on Electricity Transmission and Distribution
ADETEF	Cross-Ministry of Finance, Economy and Sustainable Development (France)
AEA	National Electrotechnical Association of Argentina
AENOR	Asociación Española de Normalización y Certificación (Spanish Association for Standardization and Certification)
AFNOR	Association Française de Normalisation (French Association for Standardization)
AG	CENELEC/CEN – Assemblage General (General Assembly)
AHG	ad hoc group
AK	DIN Arbeitskreis (task force)
AMD	amendment
AMN	American Mercosur Nations
ANAB	American National Standards Institute – American Society for Quality National Accreditation Board (United States)
Annex 7	EU European annexes on normative references to international publications
ANS	American National Standard (United States)
ANSI	American National Standards Institute
ASA	American Standards Association (United States)

ASD	ANSI Accredited Standards Developer (United States)
ASIL	Automotive Safety Integrity Level of ISO 26262
ASME	American Society of Mechanical Engineers
ASTM	American Society for Testing and Materials
BDI	Bundesverband der Deutschen Industrie (German Association of Industry)
BIS	Bureau of Indian Standards
BNQ	Bureau de Normalisation du Québec (Canada)
BS	British Standard
BSI	British Standards Institution
BSR	ANSI Board of Standards Reviewer (United States)
BT	CENELEC Technical Office (Bureau Technique)
BTTF	CENELEC Technical Board Task Force
BTWG	CENELEC Technical Board Working Group
CA	CENELEC/CEN – Committee Administrative (Administration)
CAB	IEC – Conformity Assessment Board
CACC	CEN – Committee Administrative Consulting Committee
CAE	Audit and Evaluation Committee (France, AFNOR)
CANENA	Council of Harmonization of Electrical Standardization of the Nations of America
CAS	China Association for Standardization (China)
CB-Scheme	IEC – B219Certification Bodies Scheme
CC	IEC – Compilation of Comments of Committee Draft
CCC	China Compulsory Certification (China)
CCMC	CEN-CENELEC Management Center
CCPN	Standardization Coordination and Steering Committee (France, AFNOR)
CD	IEC Committee Draft
CDV	IEC Committee Draft for Vote
CE	European Conformity
CEA	(Electrical Committee of Argentina) Comite Electrotecnico Argentino
CEI	Italian Electrotechnical Committee
CEM	Mexican Electrotechnical Committee
CEN	European Committee for Standardization (English); Comité Européen de Normalisation (French); Europäisches Komitee für Normung (German)
CEN/BT	CEN Technical Board
CENELEC	European Committee for Electrotechnical Standardization
CIF	ANSI Consumer Interest Forum (United States)
CIGRE	International Council on Large Electric Systems
CIM	Common Information Model
CIRED	International Conference on Electricity Distribution
CMC	CENELEC/CEN – Management Centre
CMF	ANSI Company Member Forum (United States)
CNAS	China National Accreditation for Conformity Assessment
CNE	National Commission of Energy (Chile)
CNIS	China National Institute of Standardization (China)
CO	IEC Central Office
COBEI	Brazilian Committee for Standardization in Electricity, Electronic, Illumination and Telecommunication

COPANT	PanAmerican Standards Commission
COR	corrigendum
CQC	China Quality Certification Centre (China)
CSA	Canadian Standardization Association
CSEE	Chinese Society of Electrical Engineering
CSIC	China Standards Information Centre (China)
CSP	China Standards Press (China)
CWA	CEN-CENELEC Working Group Agreement
DAR	Deutscher Akkreditierungsrat (Germany)
DC	direct current
DER	distributed energy resources
DGBMT	Deutsche Gesellschaft für Biomedizinische Technik im VDE
DIN	Deutsches Institut for Normung e. V.
DKE	Deutsche Kommission Elektrotechnik Elektronik Informationstechnik im DIN und VDE
DLMS	Device Language Messaging Specification
doc	CENELEC date of conformity
DoC	Department of Commerce (United States)
DoE	Department of Energy (United States)
dop	CENELEC date of publication
dor	CENELEC date of recognition
dow	CENELEC date of withdrawal
Draft prEN	CENELEC draft preliminary European norm
ECISS	European Committee for Iron and Steel Standardization
EEG	Germany – Erneuerbare Energien Gesetz (Renewable Energy Law)
EFTA	European Free Trade Association
EISA	Energy Independence and Security Act (United States)
EMC	electromagnetic compatibility
EMI	electromagnetic interferences
EN	EN:B82 European standard
FR	European normative
ENEC	European Mark for Electric Product Quality and Safety
ENTSO-E	European Network of Transmission System Operators (Europe)
EPO	European Patent Offices
EPRI	Electric Power Research Institute (United States)
ESO	European Standards Organization
ESS	European Standardization System
Essential Requirement	EU Requirement of European Union Directive on matters of safety, health or other matters covered by the New Approach Directive
ETG	Die Energietechnische Gesellschaft im VDE
ETSI	European Telecommunications Standards Institute
ETSI TC IST	Intelligent Transport Systems and Car to Car Communication
Euro NCAP	Test Procedures for Safe Cars (Europe)
FCC	Federal Communication Commission (United States)
FDA	Federal Food and Drug Administration (United States)
FDIS	IEC Final Draft International Standard
FDN	National Standardization Body (Venezuela)

FNN	Forum Netztechnik/Netzbetrieb im VDE
FprEN	CENELEC Final Draft Project European Norm
GATT	General Agreement on Tariffs and Trade
GB	National Standard (China)
GMA	Die VDI/VDE-Gesellschaft Mess- und Automatisierungstechnik
GMF	ANSI Governmental Member Forum (United States)
GMM	Die VDE/VDI-Fachgesellschaft Mikroelektronik, Mikrosystem- und Feinwerktechnik
GOST	Federal Agency on Technical Regulation and Metrology
GWAC	Gridwise Architecture Council (United States)
HBES	Home and Building Electronic Systems
HD	CENELEC – harmonized document
HV	high voltage (>1 kV)
HVDC	high-voltage direct current
IAF	International Accreditation Form
IBNORCA	Instituto Boliviano de Normalización y Calidad
ICAP	IEEE – Conformity Assessment Program
ICONTEC	Instituto Colombiano de Normas Técnicas (Colombia)
ICONTEC	Institute of Technical Standardization in Columbia
ICT	Information and Communication Technology
IEC	International Electrotechnical Commission
IEC APC	IEC Activities Promotion Committee (Japan)
IEC EE	IEC System of Conformity Assessment Schemes for Electrotechnical Equipment and Components
IEC EX	IEC System for Certification to Standards Relating to Equipment for Use in Explosive Atmospheres
IEC Q	IEC Quality Assessment System for Electronic Components
IEEE	Institute of Electrical and Electronic Engineers
IEEE-SA	IEEE – Standards Assocation
IEEJ	Institute of Electrical Engineers of Japan
IMF	International Monetary Fund
INDECOPI	National Normalization Institute of Peru
INTN	Instituto National de Tecnologia y Normalización Paraguay
IPR	International Property Rights
IRAM	Instituto Argentina de Normalización
IS	IEC International Standard
ISGT	IEEE – Integrated Smart Grid Technology
ISMS	Information Security Management System
ISO	International Organization for Standardization
ISO/CASCO	ISO – Committee on Conformity Assessment
ISO/TMB	ISO – Technical Management Board
ITA	IEC – Industry Technical Agreement
ITG	Die Informationstechnische Gesellschaft im VDE
ITU	International Telecommunication Union
JAB	Japan Accreditation Board for Conformity Assessment
JAP	Japan Accreditation Board

JAS	Japanese Standard Association (Japan)
JICA	International Cooperation Agency (Japan)
JIPDEC	Japan Institute for Promotion of Digital Economy and Community (Japan)
JIS	Japanese Industry Standards (Japan)
JISC	Japanese Industrial Standards Committee
JRCA	Japanese Registration for Certificated Auditors (Japan)
JSA MSE Dept.	JSA Management System Enhancement Department (Japan)
JTC1	ISO/IEC Joint Technical Committee
LV	low voltage (<1 kV)
MB	member body
METI	Ministry of Economy, Trade and Industry (Japan)
MoU	memorandum of understanding
MRA	Mutual Recognition Agreement (United States)
MSB	IEC Market Strategy Board
MT	IEC Maintenance Team
MultiSpeak	MultiSpeak Software Interoperability
NA Automobil	DIN Standardization of Road Vehicle Engineering (Germany)
NAFTA	North American Free Trade Agreement
NC	IEC National Committee
NCB	IEC National Certification Bodies
NEMA	National Electric Manufacturer Association (United States)
NESCOM	new standards approval committee of IEEE-SA Board
NF	Normalisation France
NIST	National Institute of Standards and Technology (United States)
NLF	EU New Legislative Framework (Europe)
NM	Norm Mercosur (South America)
NP	IEC new work item proposal
NPE	National Platform for Electromobility (Germany)
NREL	National Renewable Energy Lab (United States)
NSB	National Standards Body
NSS	National Standards Strategy for the United States
NWIP	IEC New Work Item Proposal
OAA	Organization for Argentinean Accreditation
OAS	Organization of American States
O-Member	IEC Observer member of IEC Technical Committee without voting rights
OMF	ANSI Organizational Member Forum (United States)
ONP	Oficina de Normalización Previsional (Peru)
PAR	project authorization request for new standards project or revisions
PAS	IEC – publicly available specification
PCISSC	Payment Cards Industry Security Standards Council
PES	IEEE – Power and Energy Society
PINS	Project Identification Notification (United States)
P Member	permanent member of an IEC technical committee
PNE	Présentation de Norm Européennes – rules for the structure and drafting of European Standards (PNE Rule)
PQ	CENELEC primary questionnaire to NCs

prEN	CENELEC Project European Norm
RCD	residual current device
RES	renewable energy sources
RESS	rechargeable energy storage system
REVCOM	review standards approval committee of IEEE-SA Board
RNF	Réseau Normalisation et Francophonie (France/Canada)
RR	IEC – Review Report
RvA	Raad voor Accreditatie Service (Netherlands)
RVC	IEC – Result of Voting on CDV
RVD	IEC – Report of Voting on FDIS
RVN	IEC – Result of Voting on New Work Item Proposal (NP)
SAC	Standardization Association of the People's Republic of China
SAE	International Consulting on Software and Interoperability
SAE	SAE International Engineering (United States)
SB	IEC Sector Board
SC	IEC and CENELEC – Subcommittee, CIGRE – Study Committee
SCC	Standards Council of Canada
SDO	Standard Development Organization
SEC	National Governmental Supervisory Body (Chile)
SEG	IEC – System Evaluation Group
SG	IEC – Strategic Advisory Group
SGIP	smart grid interoperability panel
SI	international system of units
SMB	IEC – Standards Management Board
SR	CENELEC – Secretary Report
Standstill obligation	CENELEC member states of EU and EFTA have to stop all standardization activities in a particular technical field until the CENELEC standardization process is concluded with an EN or with the cancellation of the work in CENELEC
StGB	Strafgesetzbuch (Germany, criminal law)
STS	Science to Standards (DKE)
SyC	system committee (IEC)
SysCom	IEC System Committee for System Orientated Standardization covering several TCs and SCs
T&D Europe	European Association of Electricity Transmission and Distribution Equipment and Service Industry
T&D	transmission and distribution
TBT	technical barriers to trade
TC	IEC, ISO and CENELEC – technical committee
TEM	transverse electromagnetic mode
TMB	ISO – Technical Management Board
TMC	Technical Management Committee (United States national committee)
TR	IEC – technical report
TS	IEC – technical specification
TTA	IEC – technology trend assessment
UA	DIN Unterausschuss (subcommittee)

UAP	CENELEC unique acceptance procedure
UAP	unique acceptance procedure for rapid approval of European standards
UCA	International User Group (United States)
UHV	Ultra High Voltage (>800 kV)
UKAS	United Kingdom Accreditation Service
UL	Underwriter Laboratories (United States)
UN	United Nations
UN ECA	United Nations Economic Commission for Africa
UN ECE	United Nations Economic Commission for Europe
UN ECLAC	United Nations Economic Commission for Latin America and the Caribbean
UN ESCAP	United Nations Economic Commission for Asia and the Pacific
UN ESCWA	United Nations Economic and Social Commission for Western Asia
UNI	Ente Italiano di Normazione (Italian Organization for Standardization)
UNIT	Instituto Uruguayo de Normas Técnicas
UQ	CENELEC – Updating Questionnaire to NCs
US NCAP	United States – New Car Assessment Programs
USASI	United States of America Standards Institute
USNC	United States National Committee to IEC
USNC Council	United States National Committee Council for IEC
USSS	United States Standards Strategy
USTAGs	United States – US Technical Advisory Groups
VDA	Automotive Industry Association (Germany)
VDE	Verband der Elektrotechnik Elektronik Informationstechnik e. V. (Germany)
Vilamoura process	CEN-CENELEC allows the preparation of a draft standard within one National Committee
W3C	World Wide Web Consortium on standards like XLM protocols
WG	IEC working group
WI	work item
WTO	World Trade Organization
WTP	Wireless Power Transfer
ZigBee	The ZigBee Alliance for Wireless Communication and Internet of Things.

Glossary

Charging modus   Process to charge the battery of an electric vehicle (IEC 61851).

Consensus   General agreement, characterized by the absence of sustained opposition to substantial issues by any important part of the concerned interests and by progress that involves seeking to take into account the views of all parties concerned and to reconcile any conflicting arguments. Note that consensus need not imply unanimity.

Consensus (ANSI)   Substantial agreement has been reached directly by materially affected parties. This signifies the concurrence of more than a simple majority but not necessarily unanimity. Consensus requires that all views and objections be considered and that an effort be made toward their resolution [1].

Consensus body (ANSI)   The group that approves the content of a standard and whose vote demonstrates evidence of consensus [1].

Continuous maintenance (ANSI)   Maintenance of a standard by consideration of recommended changes to any part of it [1].

Electromobility   The use of electric vehicles for various transportation/traffic needs.

High voltage   For electric vehicles the following definitions are used: Voltage class B, greater than 30 V AC up to 1000 V AC, or greater than 60 V DC up to 1500 V DC (ISO 6469-3). For electric power systems high voltage is defined as any voltage greater than 1 kV for alternative currents (a.c.) and 1.5 kV for direct currents (d.c.).

Periodic maintenance (ANSI)   This is defined as the maintenance of a standard by review of the entire document and action to revise or reaffirm it on a schedule not to exceed 5 years from the date of its approval [1].

Proxy (ANSI)   A written and signed document by which a voting member of a consensus body authorizes another person to vote in the member's stead, if allowed by the standard developer's procedures [1].

Resolved (ANSI)   A negative vote cast by a member of the consensus body or a comment submitted as a result of public review where the negative voter agrees to change his/her vote or the negative commenter accepts the proposed resolution of his/her comment [1].

Stabilized maintenance (ANSI)   A standard that is maintained under the stabilized maintenance option must satisfy the following eligibility criteria [1]: (i) the standard addresses mature technology or practices and, as a result, is not likely to require revision; (ii) the standard is not safety or health related; (iii) the standard has the status of an American National Standard and has been reaffirmed at least once; (iv) at least 10 years have passed since the last revision or affirmation; (v) it is used in connection with existing standards or for reference purposes.

Substantive change (ANSI)   A change that directly and materially affects the use of the standard. Examples of substantive changes: 'shall' to 'should' or 'should' to 'shall'; addition, deletion or revision of requirements; addition of mandatory compliance with referenced standards [1].

Unresolved (ANSI)   A negative vote submitted by a consensus body member or a written comment submitted by a person during public review expressing disagreement with some or all of the proposed standard, which has not been resolved satisfactorily and/or withdrawn after having been addressed according to the developer's approved procedures [1].

1

Why Standards?

1.1 General Introduction

The history of standards goes back a long way. As early as 1750 BC. the Codex Hammurabi stated : 'The master of the building will receive a death penalty if he has constructed a house which breaks down and kills the people inside' (§229). The Third Book of Moses (19: 35–36) says: 'You shall not use incorrect length, weight and volume in front of the justice. Right weight, right balance and right volumes shall be used before Jahveh, your God, who has guided you out of Egypt to meet all rules and follow the rights.' This was written around 1000 BC.

In China, in 2200 BC, the Emperor Qin Shihuangdi produced common technical standards for the wheels of transport waggons, the width of the city gates, the dimensions of the streets, for measures of length and weight and for water pipes, weapons and armour.

Standards leave room for creativity when the government of a country sets up rules for traffic or for the format of film material, or when standards define test procedures for *Bacillus cereus* on worldwide basis.

Standards leave more room for creativity where each village sets up rules for traffic or if each camera manufacturer uses his own format for the film material or if each ice-cream manufacturer defines his own test procedure for *Bacillus cereus*.

Standards leave even more room for creativity when each car driver sets up his own rules for traffic, or if each photograph uses his own film material, or each ice-cream seller tests his ice cream by watching if people feel bad after eating it, or if each computer manufacturer uses his own hardware.

Standards are everywhere but we usually do not recognize them.

Credit cards are an example of standardization. Each bank could have developed its own credit-card design. Round, square, thick, thin, one or two chips and so on. Would they have

Practical Guide to International Standardization for Electrical Engineers: Impact on Smart Grid and e-Mobility Markets, First Edition. Hermann J. Koch.
© 2016 John Wiley & Sons, Ltd. Published 2016 by John Wiley & Sons, Ltd.

gained worldwide acceptance? Which would be best for withdrawing money from machines? Even the purse has adapted to the size of the credit card.

Paper sizes (such as B4 in the United States or DIN A4 in Europe) are another example of standardization. Paper sizes are used by printers, for envelopes, publications and so on.

Nobody is obliged to use standards – the only requirement is that products are safe. Manufacturers who do not use standards may find that their products are hard to sell and that it is difficult for them to gain acceptance. Not making the use of standards mandatory opens the door for research and new developments. If all manufacturers always had to use the same manufacturing process and use the same principles, the development of new products would be hindered.

If a new credit-card design appears that is better and safer, users will accept it and a new 'standard' credit card would develop and penetrate the market. All other services using credit cards would adapt to this new standard. This is only possible if the market is flexible and standards are not mandatory.

Standards are a way to create order and give a basis for cooperation. They offer state-of-the-art solutions for continuously repeated tasks.

Standardization is a regularly planned process of writing standards according to rules. It is not carried out for the benefit of any single interested party.

Waldemar Hellmich, the first chairman of the standardization organization for general mechanical engineering, Normenausschuss der deutschen Industrie, stated in 1917: 'Writing standards is crucial work. Those involved often fight with nontechnical arguments for economic reasons.'

1.2   War on Standards

A 'war on standards' can happen when someone has an interest in avoiding standards in certain technical fields. There are various reasons why they happen. In most cases they ended up with more disadvantages than advantages for the industry and also for society. Today's leading standardization organizations follow the recommendations of the WTO, which include rules to avoid such wars on standards.

Here are some examples. The width of railroad tracks varies for different reasons – for instance competition, military reasons or strategic reasons to protect markets. In the United States the railroad track width is different in the north and south. Spain and Russia are different from the rest of Europe.

Sometimes incompatibility may exist by chance or because nobody really recognized that it would be a problem at the beginning. Once it is there, it stays for a long time. In Europe this happened with the power supply, which is different in different European countries. The so-called 'Europe plug' came much too late. Now Europe is trying to avoid making the same mistake with charging plugs for electric vehicles.

Network markets are usually dominated by one strong player. Standardization interests need to be coordinated long before technical solutions have been marketed. The goal is to produce technical solutions with compatible designs so that users are able to choose from different manufacturers.

Technologies that do not comply with standards will not develop a strong market position. They will be locked into market niches. They will either have to adapt to the mainstream or they will stop business after a while.

In more recent history a war on standards involved the digital control of factories. The so-called 'Profi Bus discussion' in the 1970s concerned a standardized bus system to control machinery in factories. Two large groups of companies fought about the definition of the design of the bus system – about how many data lines and control lines there should be and their function.

It was not possible to agree a standard bus design but technical development continues. Millions of dollars have been invested in parallel developments of two different bus systems. There was no winner. The manufacturers had large additional development and design costs and the users could not gain from the technical development, increased functionality and reduction in price that standardization might have brought. The lessons from this antistandardization fight was that none of the main players will win.

In the 1990s, when digital communication had to be standardized, the global industrial community came together to design a common standard protocol. Counterparts on both sides of the Atlantic worked smoothly together in International Electrotechnical Commission (IEC) working groups.

1.3   Main Players

1.3.1   Europe

The framework for standardization in Europe is given by the World Trade Organization (WTO) and its rules for trade without technical barriers. In the European Union, the International Organization for Standardization (ISO) produces general standards, the IEC electrical standards, and the International Telecommunication Union (ITU) telecommunication standards. Europe follows exactly the same structure, with the Comité Européen de Normalisation (CEN) for general standards, the European Committee for Electrotechnical Standardization (CENELEC) for electrical standards and ETSI (European Telecommunications Standards Institute) for telecommunication standards – see Table 1.1.

This structure makes the European Union a very homogeneous region with regard to standards. The main goal of the member states was to create free market access for all member countries. The 28 member countries following EN standards make it the largest single market place in the world with more than 500 million people from Norway to Sicily and from Portugal to the Baltic countries.

Before harmonization, in the electrical field in different countries of the European Union, a total of close to 30 000 standards were used. Today the number is down to about 6000, including standards for new technologies that did not exist before.

The basis for this harmonization was organized by new European institutions – CEN, CENELEC and ETSI – with the technical support of experts financed by the industry.

Table 1.1   European standardization organizations.

	Europe	International
General	CEN	ISO
Electrical	CENELEC	IEC
Telecommunication	ETSI	ITU

No government was involved directly and this remains the case today. These so-called 'self-regulating' bodies are purely focused on technical questions and not on political issues. The European standardization organizations are, in principle, financially independent from politics and industry because their main finances come from membership fees of national standardization organizations, which generate their main income by selling standards.

In real life nothing is independent of political issues or social trends but CEN, CENELEC and ETSI can determine their own direction. European nations sometimes set up rules that are in conflict with EN standards. In such cases, so-called deviations are used for particular countries. Deviations from EN standards are possible but have to be harmonized as soon as possible.

For example, pressure-vessel applications have to follow national regulations or laws because politicians want to protect their people from exploding vessels. That is clear and understandable. On the other hand, different test methods and calculation rules require expensive development processes and testing. If each country has its own set of rules, the result in the end in each European country is the same: pressure vessels are safe – but the way in which this has been achieved is different in each country.

Gas-insulated substations contain devices to switch high-voltage power lines; the enclosures are pressure vessels and have to follow national rules. Before European harmonization the rules were different in almost any country. If a manufacturer placed an offer he had to calculate the additional cost for testing and certification. As all manufacturers must do so there is no competition and additional costs go directly to the customer and then to the electricity consumer. Before harmonization in Europe took place there were more than 20 different requirements for such pressure vessel tests and certification.

Today only one requirement for pressure vessels by a related EN standard is in place and only one deviation is left in the EN standards for high-voltage switchgear assemblies, which is for Italy. All other national legislatures have changed and adapted their regulations or laws for pressure vessels to the EN standards. For sure Italy will follow soon to adapt their national regulations. The advantage is clear. Today the manufacturer needs to follow only one procedure for testing and certification of pressure vessels for high voltage switchgear assemblies and he can serve all the EU member countries, only Italy requires special design rules.

Today the same EN standards are widely accepted outside Europe – in America, Africa, Asia, and Australia.

The European Union is often seen as overregulating daily life. Examples of this are requirements concerning the size of bananas, the radius of cucumbers, or the size of steps on ladders. This might be true and such requirements might be unnecessary but in the technical field of electrical equipment, services and systems, standardization in Europe has provided large benefits for users. In many cases European standardization is relevant globally because of active European participation in global standardization organizations like the IEC, ISO and ITU.

Standardization in Europe is often driven by industry. This does not only involve large industrial players; in many cases small and medium-size companies bring their knowledge and look for a global market. Small and medium-size companies use international standardization to make their products available on a global basis without needing sales and technical offices in any country.

International standardization promotes innovation and new technical solutions in a global market, gives new opportunities and helps to spread products in the world market.

1.3.2   America

Standardization in America differs from Europe on one key point: there are many standardization organizations providing standards at a local or regional level or in technical branch. In the electrical field the main organizations are the IEEE and the IEC; in some cases EN standards are also applied. Many sector-related organizations offer their standards in the market.

The Pan-American Standards Commission (COPANT) is a civil, nonprofit association. It is financed by its membership fees and by grants from the Organization of American States (OAS). The key objectives are the promotion of the development of technical standardization in its member countries to evaluate existing standards and to resist attempts to develop national standards if international or regional standards meet COPANT's national requirements.

The Forum of IEC National Committees of the Americas (FINCA) is a coordination body founded 2007 in Ottawa, Canada. The member countries coordinate their standardization interests every year at an IEC General Meeting. The members are Canada, the United States, Mexico, Colombia, Brazil, Argentina and Chile. The activities of American nations in the IEC and ISO are shown in Tables 1.2 and 1.3.

The dominant IEC member countries are the United States and Canada, which have a high level of participation in IEC technical committees (TCs) and hold many leading positions on the IEC as chairmen and secretaries. Mexico and Brazil are building up active participation in the IEC, while Argentina, Columbia, Cuba and Chile are just starting their activities with the IEC.

The long list of ISO members in America shows the wide spread of this international organization and its deep links with even the very small nations. Many more member states are active with the ISO than the IEC.

The Council of Harmonization of Electrical Standardization of the Nations of America (CANENA) is a regional standardization organization in North America's NAFTA region. It was founded in 1992 to foster the harmonization of electrotechnical standards, conformity assessment test requirements and electrical codes. It does not publish its own standards but coordinates participation in standard-writing organizations.

There is a similar organization in the Mercosur countries of South America. American Mercosur Nations (AMN) coordinate standardization activities with the goal of harmonization. It does not publish standards.

Table 1.2   American members in IEC technical committees (TC). The numbers show how many experts have been nominated to the organizations.

Nation	Organization	P member of TC	O member of TC
Argentina	CEA	4	11
Brazil	COBEI	31	51
Canada	SCC	93	26
Mexico	CEM	46	57
USA	ANSI	154	–
Colombia (associated)	ICONTEC	4	–
Cuba (associated)	ICONTEC	4	–
Chile (in preparation)			–

Notes: P – member has full voting rights; O – Member can comment.

Table 1.3   American members on ISO technical committees.

Nation	Organization	Nation	Organization
Argentina	IRAM	Jamaica	BSJ
Barbados	BNSI	Mexico	DGN
Bolivia	IBNORCA	Nicaragua	SON
Brazil	ABNT	Panama	COPANIT
Canada	SCC	Paraguay	INTN
Chile	INN	Peru	INDECOPT
Colombia	ICONTEC	Saint Lucia	SCBS
Costa Rica	INTECO	Saint Vincent and the Grenadines	SVGBS
Cuba	NC	Suriname	SSB
Ecuador	INEN	Togo	CSN
El Salvador	CONACYT	Trinidad and Tobago	TTBS
Guatemala	COGUANOR	United States	ANSI
Guyana	GNBS	Uruguay	UNIT
Honduras	COHCIT		

Table 1.4 provides an overview of American organizations involved in standardization at a national level.

In South America there is some competition between standardization organizations. On one hand, organizations from the NAFTA area in North America, like the American National Standards Institute (ANSI), the Institute of Electrical and Electronic Engineers (IEEE), Underwriter Laboratories (UL) or the National Electric Manufacturer Association (NEMA), are promoting their standards; on the other hand, the international standardization organization, IEC, based in Europe, offers its own standards.

In Brazil, Argentina, Colombia and other nations, this results in requirements that specify several standards, so devices and products are overdesigned and more expensive. This situation also leads to higher costs for design, development and manufacturing because of smaller numbers following each standard. At the end this leads to higher costs for users and customers.

Such conflicting standardization can also be seen in South Africa, Australia and parts of Asia. In recent years a new player in the field of standardization has arrived on the scene: China and its SAC standards.

1.3.2.1   United States of America

1.3.2.1.1   General

The electrical network in the United States is one of the largest in the world and so is the standardization structure and organization. Since early industrialization, the United States has been continuously developing the electric system and technical standardization has developed with it.

Spread over many industrial sectors, business is a big driver for standardization. Once a new business opportunity is identified the related standardization organizations develop new standards in competition and government organizations support standardization with laws, regulations and, in some cases, with grants to finance expert groups.

Table 1.4 An overview of American organizations involved in standardization at a national level.

Country	Organization	Abbreviation
Argentina	Instituto Argentino de Normalización y Certificación	IRAM
Barbados	Barbados National Standards Institution	BNSI
Bolivia	Instituto Boliviano de Normalización y Calidad	IBNORCA
Brazil	Asociación Brasileira de Normas Técnicas	ABNT
Canada	Standards Council of Canada	SCC
Chile	Instituto Nacional de Normalización	INN
Colombia	Instituto Colombiano de Normas Técnicas y Certificación	ICONTEC
Costa Rica	Instituto de Normas Técnicas de Costa Rica	INTECO
Cuba	Oficina Nacional de Normalización	NC
Ecuador	Instituto Ecuatoriano de Normalización	INEN
El Salvador	Consejo Nacional de Ciencia y Tecnología	CONACYT
Grenada	Grenada Bureau of Standards	GDBS
Guatemala	Comisión Guatemalteca de Normas	COGUANOR
Guyana	Guyana National Bureau of Standards	GNBS
Honduras	Consejo Hondureño de Ciencia y Tecnología	COHCIT
Jamaica	Bureau of Standards Jamaica	BSJ
México	Dirección General de Normas/Secretaría de Economía	DGN
Nicaragua	Ministerio de Fomento, Industria y Comercio y Dirección de Tecnología, Normalización y Metrología	MIFIC
Panamá	Comisión Panameña de Normas Industriales y Técnicas	COPANIT
Paraguay	Instituto Nacional de Tecnología y Normalización	INTN
Perú	Instituto Nacional de defensa de La Competencia y de La Protección de La Propiedad Intelectual	INDECOPI
República Dominicana	Dirección General de Normas y Sistemas de Calidad	DIGENOR
Santa Lucia	Saint Lucia Bureau of Standards	SLBS
Trinidad y Tobago	Trinidad and Tobago Bureau of Standards	TTBS
United States	Estados Unidos de América American National Standards Institute	ANSI
Uruguay	Instituto Uruguayo de Normas Técnicas	UNIT
Venezuela	Fondo Para la Normalización y la Certificación de la Calidad Fdn: Fondo de desarrollo Para la Normalización, Calidad, Certificación y Metrología	FONDONORMA/FDN

Smart Grid is an example of how the standardization process works in a new technical field in the United States. It will bring new intelligent sensors, meters and digital control devices into use and will create solutions for a complex electric power system. Electric power generation in the Smart Grid world will use a mixture of generation resources – gas, coal nuclear, wind, solar, geothermic, biomass and water. In addition, different storage technologies for electricity will be developed with hydropower storage plants, chemical processes to generate hydrogen from wind or solar energy and electric batteries. Combined with intelligent electric loads for industrial processes, home electric management systems, electromobility and decentralized generation with photovoltaic panels on the roof, the need for digital real-time control offers many new business models for saving, selling or trading electricity. All these activities will drive new business in the electricity market and need standardization.

Standard-developing organizations combine their industrial activities in different subject-related organizations under competitive conditions. The National Electric Manufacturer Association (NEMA) has created a Smart Grid Advisory Panel to formulate requirements on technical standards for Smart Grid.

The US IEC National Committee, with the support of NEMA, formulated the Gridwise Architecture Council (GWAC) to present this to IEC TC 8, to initiate the Smart Grid Strategic Advisory Group in the IEC to study the role of the IEC in international standardization efforts for Smart Grid. The US National Committee was selected to lead this effort for the United States.

In the IEEE Standards Coordination Committee, SCC 21 conferences have been hosted by IEEE on the issue of Smart Grid and to develop documents like Energy 2030 or IEEE P2030 to gain a lead in the technical interpretation of the influences on, and needs of standardization for, Smart Grid.

The IEEE played a central role in contacting the National Renewable Energy Lab (NREL) and the Department of Energy (DoE). From this, the IEEE Standards Association Board started several standardization projects in different technical areas on Smart Grid.

This short example should show how standardization organizations become involved as soon as industry has identified future business. Without future business no standardization work will be started because the experts in the standardization working groups are from industry and are getting paid by industry. Finishing a standard typically requires a period of 3–4 years with two or more experts' meetings in the United States or internationally.

How is this environment structured in the United States?

1.3.2.1.2 Legal Framework
The responsibility for standardization on the government side is linked to the Department of Commerce (DoC). The National Institute of Standards and Technology (NIST) was established based on the National Technology Transfer and Advancement Act of 1996. The NIST signed a Memorandum of Understanding (MoU) with the American National Standards Institute (ANSI) to put in place rules about how standards have to be written. It received the right to control the standard-writing process for any accredited standard-developing organization in the United States (see Figure 1.1).

The US regulator uses voluntary standards to offset the need for additional regulations or to enhance existing regulations. On the other hand, if regulations are necessary, US regulators are required by law to use voluntary standards whenever possible. There are more than 6500 laws and regulations in place that refer to standards.

1.3.2.1.3 Comparison between the United States and the European Union
The legal framework in the United States is different from the situation in Europe, where European Directives only provide the framework for standards. Standards in Europe (EN) are the technical specifications for the directives.

Figure 1.1 Legal framework of standardization in the United States.

In the United States, standardization is seen as national, whereas Europe operates on a multinational basis. Each European standard must be changed into a national standard and must remain in accordance with national laws and regulations.

In the United States, standards are primarily seen as strengthening the economy and improving quality of life. In Europe the primary goal of EN standards is to harmonize national conditions and to reduce trade barriers in the European Union.

In the United States consensus standards come from industry or governmental agencies. In Europe there are only two consensus standard organizations: CENELEC for electricity and CEN for anything else. The United States has more than 270 standardization organizations accredited by ANSI – agency standards, for example FCC or FDA, and standards from NIST.

In Europe only three standardization organizations release European standards: CEN, CENELEC and ETSI. In addition, each of the 28 member countries has its own standardization organization, which has to follow EN standards when they are available and can write its own standards if no other EU country is interested – see Chapter 4.

In the United States the application of standards is voluntary in general and only required in the regulated domain of public entities. In Europe, the EN standards are only applied voluntarily. In cases when European Directives (regulations) apply, standards may be mandatory. In such cases the manufacturer or trader displays the EU conformity sign to indicate compliance with these standards and regulations.

In the United States, national and international standards coexist and might conflict. Market participants are free to choose the standards that best fit their needs. It is not the case that 'one size fits all'. In Europe, CEN and CENELEC have formal agreements with ISO and IEC (the Vienna and Dresden Agreements) to reach the maximum level of convergence – see Table 1.5.

The agreement with the World Trade Organization (WTO) in the United States is signed by ANSI, which is linked to the government. In Europe the WTO agreements are signed by CEN, CENELEC and ETSI, which are private organizations with a nonprofit orientation and no direct government control.

1.3.2.1.4 Understanding the US Standardization System

There are several misunderstandings about the standardization system in the United States. The European view of the United States is often influenced excessively by the systems used in Europe and does not see that the different approach in the United States also works very efficiently.

Table 1.5   A comparison of US and EU standardization.

United States	European Union
National	International
Strengthen economy and improve quality of life	Remove national trade barriers
• Industry consensus	Only CEN and CENELEC consensus
• Governmental agency consensus	
• 270 SDOs accredited by ANSI	• CEN, CENELEC, ETSI
• Agencies (FCC, FDA, NIST)	• National Institutes
Voluntary application when unregulated	Voluntary application with presumed conformity
National and international standards coexist	Formal agreements
	ISO-CEN and IEC-CENELEC
WTO signed by ANSI	WTO signed by CEN, CENELEC, ETSI

One statement that is often heard is that the United States is not committed to ISO or IEC. This is not true. The US National Committee, ANSI, is one of the biggest contributors to ISO and IEC. This is the case for leading roles like chairman and secretaries and also at the level of experts in the working groups. In some technical fields, like digital communication or computers, the United States sometimes has a dominant role.

What the United States does not do is to adopt any ISO or IEC standard routinely in a national standard as is done in Europe with the EN standards – with a requirement that it be adopted by national standardization organizations of the members of the EU. On the other hand this does not mean that ISO or IEC standards are not used in the United States. The use of IEC and ISO standards is up to the standard user. Users of standards are free to choose those standards that best fit their needs. Only in cases where there are regulations or legal requirements in the United States might it be necessary to follow specific standards, such as the National Safety Code. So standard users in the United States are much freer to use standards they like most.

Another statement often heard is that the United States is not committed to the withdrawal of conflicting standards. The use of ISO and IEC standards is not mandatory in the United States. There is no centralized organization, like CENELEC in Europe, to withdraw standards of other organizations. The decision to use an ISO or IEC standard is made by the standard user and depends on his requirements.

A further statement that is heard is that the United States does not use ISO or IEC standards. This is not true. The United States is using ISO and IEC standards but the United States believes that standards from other long-established organizations that develop and publish standards have a role that complements that of the ISO and IEC in certain technical fields. Again it is up to the user of standards to decide which standard his specification will follow. This could also be a mix – one requirement from one standard and another requirement from another standard.

Some believe that the United States, with ANSI, promotes standards that originate in the United States ahead of ISO and IEC standards. This is not correct: ANSI promotes the use of market-relevant international standards whoever publishes them and wherever they are published. They may originate from the ISO or the IEC or another standardization organization. What is more important is the standard user's experiences with standards and his specific knowledge of the technical details of standards. If some company has had good experiences with specifications that follow, for example, an IEEE standard, it is very likely that the next specification will follow the related IEEE standard too. In such cases it will be not be easy to convince the user to use any other standard, such as IEC or ISO standards.

The statement that the standardization system in the United States is fragmented, with many uncoordinated standardization organizations, is not really true. At the first glance the United States has more than 270 accredited standard-writing organizations, which is certainly a big number. But looking more deeply, many organizations are working in different technical areas or operate in different regions or have different organizational structures (private, public, state, county, city). The standardization system of the United States is more decentralized but it is not fragmented. Real duplication of work and standards is prevented by the well coordinated voluntary expert system, with experts sent and paid by companies. They will not pay if a standard that they need has been published somewhere – they just will use it.

It is true that the United States has differing state requirements on technical standards. This is the case because the US constitution grants this freedom to any state in the federal union. Different state requirements lead to different effects in technical sectors and ends up with different strategies. One significant example today is that Texas focuses on the installation of renewable wind energy generation whereas Arizona focuses on solar photovoltaic generation. Different strategies are leading to different state tax and sponsoring models, which lead to different results in the end. This is often not understood in Europe, which sees the United States as one nation. In many aspects state rules differ widely from state to state – as in European nations.

1.3.2.1.5 *Standardization Entities*

In the United States a large number of entities are engaged in the standardization market. Driven by market interests and future business possibilities, industries and experts are entering the standardization scene. In most cases the experts are contributing to one of the 270 accredited standardization organizations or they start with a new consortium – whichever best fits their needs. Government has responsibilities for controlling the standardization market to maintain high-quality standardization organizations.

The roles of the main players in the United States are discussed below.

Government

The Department of Commerce (DoC) is responsible for national and international standardization. Technical work on standardization in the United States is delegated to NIST and to ANSI.

The government also provides financial support to ANSI to fulfil its duties and to set up specific technical investigations to promote and prepare standardization.

Outside the United States, embassies support standardization. United States' embassies operate so-called standards attachés to represent the United States in standard activities in countries where they are based. This shows how important standardization is regarded by the US government.

In bilateral agreements with other nations the United States supports the World Trade Organization (WTO) in removing trade barriers. In such agreements, trade rules between the nations are linked to the use of standards following the rules of the WTO.

The US government also gives direct financial support to the outreach programs of standard-developing organizations in important regions. With such financial support, technical experts are sponsored, for example, to participate in technical standardization meetings worldwide. This financial support shows the importance of international trade for US industry.

Examples

These examples are taken from the Department of Commerce's Market Development Cooperator Program – web site: http://trade.gov/mdcp/(accessed 2 March 2016).

• Steel Founder's Society of America. Led by this trade association, a Market Development Cooperator Program award of $172 000 will support experts' participation in ISOTC 17 for steel products with the intention of influencing European and Chinese standards.
• Consortium for Standards and Conformity Assessment (CSCA). A cooperative effort of standards organizations American Society of Mechanical Engineers (ASME) and the ASTM: American Society for Testing and Materials (ASTM), the American Petroleum Institute (API) and CSCA America. The $ 400 000 MDCP award and technical assistance will help CSCA establish a presence in Beijing. The goal of CSCA Beijing is to form relationships with peer agencies, monitor standards development and promote acceptance of consortium members' standards and conformity assessment systems.

These two examples show how financial support is organized in a very transparent way by making it public on the Internet, together with the goals linked to the money spent.

National Institute of Standards and Technology (NIST)

The National Institute of Standards and Technology (NIST) was founded in 1991 as a nonregulatory federal entity within the US Department of Commerce for Technology Administration. The mission of NIST is to develop and promote measurements, standards and technology and to enhance productivity, facilitate trade and improve quality of life. It does not produce and publish standards or regulations. There is no formal relationship or mandate between NIST and ANSI. More about NIST can be found under http://www.nist.gov/(accessed 2 March 2016).

The American National Standards Institute (ANSI)

The American National Standards Institute (ANSI) plays a central role in the standardization system of the United States. It controls US participation in ISO and IEC as one of its major roles. An industry, as a member of ANSI, nominates technical experts to be sent to standardization organizations. For IEC, ANSI forms the United States National Committee (IEC US NC) as the formal link to IEC. Experts are nominated and standards are voted on and commented on. The US NC of the IEC sends delegates to represent the United States at technical meetings.

Technical experts are nominated by ANSI to the organization levels of ISO at working level and at leaders' level in technical committees.

The American National Standards Institute represents the United States in the International Accreditation Forum, a worldwide organization for coordinating conformity assessment accreditation bodies in the field of management systems, products, services, personnel and other similar programs of conformity assessment (www.iaf.nu, accessed 7 February 2016).

It also sends delegates to COPANT, the pan-American organization for standards coordination, and to PASC, the Pacific Area Standards Congress, which coordinates standardization

---

Box 1.1   ANSI-accredited US standard-developing organizations.

3-A	ASC X9	ASA	ACCA	AMCA	ARI	ATIS	AA	AAMA	AAMVA
ABMA	ABYC	ABMA	ACC	ACI	ADA	AFPA	AGA	AGMA	AH&LA
AIHA	AIAA	AISC	AITC	AISI	ALI	ANS	ANLA	API	ASNT
ASQ	ASAE	ASB	ASCE	ASHRAE	ASME	ASSE	AWWA	AWS	AWEA
ATA	ACMI	ASIS	AIIM	AMT	NPES	AAMI	ACDE	AHAM	ARMA
ASTM	AIM	AGRSS	ALI	BHMA	BICSI	BOMA	BIFMA	CCPA	CSAA
CAPA	CLSI	CFPMI	CAP	CPA	CAGI	CGA	CAM-I	CEA	CSPA
CEMA	CTGI	CSA	DISA	DASMA	EIMA	EASA	EIA	ESTA	EIA
EOS/ESC	FCI	FM	GTEEMC	GICC	GEIA	BEI	HPVA	HIBCC	HL7
HPS	HFES	HI	IESNA	ITSDF	IEEE	IEST	IIE	INMM	I2AMA
IAF	IAAMC	IAPMO	ICPA	ICC	ITI	NETA	I3A	IIAR	ISEA
ISO	ASANTA	IWCA	IPC	ISA	JCSEE	KCMA	LIA	MSS	MHI
MBC	NACE	NAHBRC	NAAMM	NBBPVI	NBFAA	NCMA	NCSL	NCPDP	NECA
NEMA	NFPA	NGA	NGCMA	NISO	NIMS	NIST/ ITL	NPPC	NSC	NSAA
NADCA	NERC	NAESB	NALFA	NASPO	NSF	NIRMA	OLA	OPCC	OEOSC
OPEI	PMMI	PSA	PCA	PWMA	PMI	RPTIA	RSTC	RVIA	RESNA
RIA	RMA	SIA	SSFI	SIA	SMA	SPRI	SBS	SAE	SCTE
SMPTE	SVIA	SAAMI	SES	SDI	SJI	SSCI	TIA	TCATA	CI
TMS	SPI	TCA	TOY-TIA	TAPS	TCIA	TPI	USDA	USPRO	UL
UAMA	UAMA	UCC	VITA	WQA	WDMA	ECMA	WMMA		

in the Pacific Rim countries. In the United States today there are more than 270 standard-developing organizations accredited by ANSI. Box 1.1 shows a sample of 188.

The American National Standards Institute has an annual certification process in place to control the standardization work of accredited standards organizations.

To express the role of ANSI in a more striking form: ANSI is preparing the rules by which standard-developers operate, driven by experts from companies, government, agencies and other standard users.

In some regulated market environments, ANSI and industry cooperate closely with regulatory bodies like the Federal Communication Commission (FCC) for standards involving, for example, radio frequencies, or with the Federal Food and Drug Administration, for standards in the health sector.

International Electrotechnical Commission participation is the responsibility of the US National Committee Council (US NC Council). The members of the US NC Council are nominated by the board of directors of ANSI. The US NC Council nominates experts to working groups or maintenance teams in the IEC or it delegates representatives of the United States for the Technical Committee meetings. The US NC Technical Management Committee also administers the IEC secretariats and chair positions held by the United States in the IEC.

The US NC IEC Council coordinates the involvement of the United States in the IEC and other electrotechnical bodies associated with the IEC. It also coordinates US NC activities with appropriate standards boards to promote consistency between those international and national activities that fall within the scope of the IEC.

1.3.2.1.6 American Society of Mechanical Engineers

The American Society of Mechanical Engineers (ASME) is a globally recognized standardization organization used in more than 100 countries. There are 24 ASME Codes and Standards Programs, with standards translated into six different languages.

The ASME standards mainly cover boilers and pressure vessel requirements. ASME also prepares rules for the accreditation process. Such boilers are widely used in power plants to heat up the water to steam and use it in steam turbines to generate electricity – an important technical field. Others are used in chemical plants, refineries, pharmacy, pipelines and many other applications with pressure in an enclosure. These ASME codes and standards are widely used in North, Central and South America, in Asia and Africa but not so much in Europe where the so-called Euro Code is used.

Other top-ranking technical areas for the application of ASME are piping for gas, water and liquids; elevators; escalators; nuclear codes and the performance of test codes.

1.3.2.1.7 American Society for Testing and Materials

The American Society for Testing and Materials (ASTM) was founded in 1898 in the United States. It has developed and published more than 14 000 standards based on an international, voluntary consensus process. More than 30 000 experts from over 125 countries are participating in the standardization work and 138 technical committees focus on areas such as materials, products, systems and services, and develop standards for technical fields, as shown in Box 1.2.

The international impact of ASTM may be seen from the fact that 40% of its standards are sold outside the United States. There are memorandums of understanding with national standard bodies in over 40 nations for the use of ASTM standards.

ASTM is the number one standard-developing organization in the United States on the government regulatory list. This means that more than 3000 standards out of a total of 6500 standards are referenced by the US Codes of Federal Regulations.

1.3.2.1.8 Underwriters Laboratories

Underwriters Laboratories (UL) is a third-party certifier. In recent decades UL developed into a more global organization, with 25 offices around the world and with an increased amount of work done outside the United States.

The mutual recognition agreement (MRA) with Canada is the basis for accepting test reports and certification. This has the advantage that only one witnessing test is required to allow joint submission to both the UL for the United States and the CSA for Canada. The Component Recognition Program provides acceptance by one organization of certification by the other organization.

Box 1.2   Technical fields of ASTM.

Metals	Paints	Plastics
Textiles	Petroleum	Construction
Aviation	Energy	Medial services and devices
Consumer products	Environment	Computerized systems
Electronics	Homeland security	

1.3.2.1.9 Impact on the World Market

The strategic goal of US standardization organizations to become active on the world market with their own standards beside ISO and IEC will lead to a market split. The market will be divided, with ISO/IEC versions and US-based standards and different requirements. This is a disadvantage for the user or consumer because the scale of the market shrinks; the manufacturer needs to provide different technical solutions at higher development and design costs.

The main areas of conflict today are in Brazil and China, where both standardization worlds exist side by side and in conflict. Global harmonization and coordination is necessary to reduce high costs and to save the limited time of experts involved in writing standards. It is not possible for manufacturers and users to work in parallel in different standardization organizations on the same subject.

It will be to the benefit of all who are involved with international organizations to promote only harmonized standards coordinated by technical fields. The convergence of standards is important; overlap is expensive and hinders technical development. A fight on standardization content of standards must be avoided for the sake of all participants. Cooperation is needed.

The technical field of standardization is growing so it is necessary to share work between all standard-writing organizations.

1.3.2.2 Mercosur

Mercosur is a union of nations in South America, except the Andean nations. Mercosur nations are Argentina, Brazil, Paraguay, Uruguay and, as invited members, Bolivia and Chile. The Mercosur nations are organized in the standard-developing organization, Mercosur Association of Standardization (AMN). The members are shown in Table 1.6.

The organization of AMN is shown in Figure 1.2.

The members of AMN are delegated by the national committees of the member countries. They are delegates of the Directive Council and the Executive Secretariat. The standardization areas are split into technical committees (CSM) where CSM 01 is responsible for electricity and CSM 03 for electronics and telecommunication. The electricity sector is further split into four subcommittees (SCM), where SCM1 is responsible for electricity, SCM2 for cables and conductors, SCM3 for consumer goods and SCM4 for lighting.

The development of standards in AMN occurs in the following steps. At the draft stage, the document is sent to the members for consideration and comments. After final draft document is reached, an approval process by consensus is started. The document will be sent to national members of the technical committee (CSM or SCM) for balloting and comments. When the national standardization bodies have approved the document it will be published and registered as a Mercosur standard.

There are more than 30 technical committees for different sectors of standardization as shown in Table 1.7.

One principle of standardization in AMN for Mercosur is the prioritization of international standards coming from IEC, ISO or ITU. If no standard is found there, the priority is next given to the regional standards of COPANT for all South America. Only if no standard can be found at the international level and the regional level of South America will AMN publish a new document for the Mercosur nations.

Table 1.6   Members of Mercosur Association of Standardization (AMN).

Nation	Organization	Founded	Type
Argentina	IRAM Instituto Argentina de Normalización www.iram.org.ar (accessed 11 February 2016) Private organization	1935	Full member
Brazil	ABNT Associação Brasileira de Normas Técnicas www.abnt.org.br (accessed 11 February 2016) Private organization	1940	Full member
Paraguay	INTN Instituto National de Tecnologia y Normalización Paraguay www.intn.gov.py (accessed 11 February 2016) Governmental organization	1963	Full member
Uruguay	UNIT Instituto Uruguayo de Normas Técnicas www.unit.org.uy (accessed 11 February 2016) Private organization	1939	Full member
Bolivia	IBNORCA Instituto Boliviano de Normalización y Calidad www.boliviacomercio.org.bo/ibnorca (accessed 11 February 2016) Private organization	1963	Invited member
Chile	INN Instituto Nacional de Normalización www.inn.cl (accessed 11 February 2016) Governmental organization	1962	Invited member

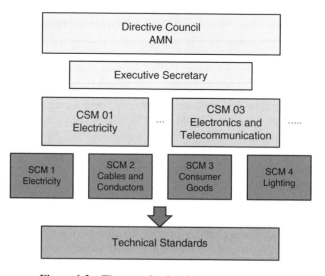

Figure 1.2   The organizational structure of AMN.

Table 1.7 Sectors of CSM technical committees.

CSM No.	Sector	Secretariat
CSM 01	Electricity	ABNT, Brazil
CSM 02	Metallurgy	IRAM, Argentina
CSM 03	Electronics and telecommunication	ABNT, Brazil
CSM 04	Toys	IRAM, Argentina
CSM 05	Cement and concrete	ABNT, Brazil
CSM 06	Mechanical machines and equipment	ABNT, Brazil
CSM 07	Automotive	IRAM, Argentina
CSM 09	Plastics for civil construction	ABNT, Brazil
CSM 12	Paper and cellulose	ABNT, Brazil
CSM 13	Quality	IRAM, Argentina
CSM 16	Environmental management	ABNT, Brazil
CSM 17	Accessibility	UNIT, Uruguay
CSM 18	Graphic technology	ABNT, Brazil
CSM 20	Clinical analysis and *in vitro* diagnosis	ABNT, Brazil
CSM 21	Sheet glass	ABNT, Brazil
CSM 22	Conformity assessment	IRAM, Argentina
CSM 23	Tourism	ABNT, Brazil
CSM 24	Nondestructive testing	IRAM, Argentina
CSM 25	Packaging	INTN, Paraguay
CSM 26	Food safety	IRAM, Argentina
CSM 27	Software quality	UNIT, Uruguay
CSM 28	Information safety	ABNT, Brazil
CSM 90:01	Lighters	IRAM, Argentina
CSM 90:02	Gas cylinders	IRAM, Argentina
CSM 90:03	Vehicles using gaseous fuels	IRAM, Argentina
CSM 90:04	Textiles	ABNT, Brazil

The European standards of CEN and CENELEC will also be checked before ANM starts its own standardization work. To coordinate these activities AMN has reached agreements with ISO, IEC, COPANT, CEN and CENELEC.

The goal of this coordination activity of AMN is to reduce deviations in different standards as much as possible for the Mercosur market. This is to the benefit of users and consumers, to avoid additional costs arising from uncoordinated standard requirements. Differing standard requirements are trade barriers between the Mercosur countries and the rest of the world. National deviations in standard requirements should be avoided.

Cooperation between Brazil, Argentina and Chile on the one hand and CEN and CENELEC on the other hand in recent years, based on a cooperation protocol signed in 2004, and stronger engagement in IEC and ISO standards, could improve standardization harmonization in Mercosur nations.

Publications

Publications in Mercosur nations are released by AMN and are identified as Norma Mercosur (NM) with a number. They are published by the national standardization bodies. There are three types of NM publications: two are based on international standards (ISO and IEC) and the other is made by AMN.

International Standards without Deviations

In the case of a Mercosur publication based on an international standard, the identification is as in the following example:

NM-ISO 15189:2008
Laboratorios de análisis clínicos.
Requisitos particulares para la calidad y la competencia.
Requisitos especiáis de qualidade e competencia.
Sustituyendo a la norma IRAM-ISO 15189:2005

International Standard with Deviation

In the case of a Mercosur publication based on an international standard the identification is as in the following example:

NM 60884-1:2001 (IEC 60884-1:1994, MOD)
Plugues e Tomadas – Uso Doméstico e Análogo, ABNT NBR

American Mercosur Nations (AMN) Standard for Mercosur

In cases where there is no international based standardization publication the identification is as in the following example:

NM 243:2000
E + L Series – Nora Lighting

1.3.2.2.1 *Argentina*

The standardization organization in Argentina is embedded in the Mercosur regional structure and international standardization. Figure 1.3 shows the structure.

The standardization process in Argentina is led by IRAM for the nomination of experts to the Mercosur standards organization (AMN) and to the ISO. The National Standardization Committee of Argentina (CEA) is the link to the IEC. It participates in technical work by nominating experts or sending comments and voting on IEC documents. The National

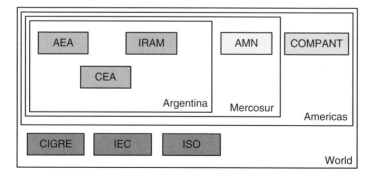

Figure 1.3   Structure of standardization in Argentina.

Table 1.8 Accredited product certifiers.

Organization	Area of activity
IRAM Instituto Argentino de Normalizacion y Certificacion	Electric, electronic toys, personal protection, auto parts, bicycles, lightners, latex painting, refrigerator, energy labelling, coolers, lamps, fruit, vegetables, flowers, coffee, tea
Bureau Veritas Argentina S.A.	Electric products, lightners, energy efficiency, refrigerators, coolers, efficiency of lamps
TÜV Rheinland Argentina S.A.	Electric products, steel, toys, energy efficiency, labelling for refrigerators, lamps, air conditioners, wind turbines
IQCSA	Electric products, toys, lightners, energy efficiency, labelling for lamps and EPP
UL de Argentina S.R.L.	Electric products for personal protection
IRAM-ITS-IMQ S.A.	Electric products
Instituto del Gas Argentina S.A.	Electric products
Net Connection International S.R.L.	Electric products
CAYLAS Consultores Assesiados S.R.L.	Electric products
SGS Argentina S.A.	Fruit and vegetables

Electrotechnical Association of Argentina (AEA) participates in the international work of the International Council on Large Electric Systems (CIGRE).

Accreditation in Argentina

There are several local and international organizations active in Argentina for Product Accreditation and Certification. Table 1.8 gives some examples.

Standardization System in Argentina

The national standardization organization of Argentina, which represents the country in the IEC, is Comité Electrotécnico Argentina (CEA). The AEA is responsible for standardization of electrotechnical installations including the certification of electrical installation personnel and companies. The standardization of products, services, systems and processes in Argentina is the responsibility of IRAM.

The different industrial sectors are organized in associations covering the technical matters in their related field. Some examples include CACAAU, CAEM, CAMARCO, and CALAMA. Test laboratories in Argentina are LENOR, INIT or UT and the active certifiers are listed in Table 1.8.

Argentinean Electrotechnical Committee (CEA)

The CEA was founded in 2001 as the official representative of Argentina in the IEC. It consists of IRAM, the national standardization and certification organization, and the AEA. The primary work of IRAM is in the area of product standardization and that of the AEA is in the area of electrical installation standardization. The CEA has established internal committees, mirroring the technical committees of IEC as shown in Table 1.9.

Table 1.9 Mirror committees of the IEC in Argentina.

Committee	Title
TC 14	Power transformers
SC 23E	Circuit breakers and similar equipment for household use
TC 42	High-voltage testing techniques
TC 45	Nuclear instrumentation
TC 57	Power system management and associated information exchange
TC 61	Safety of household and similar electrical appliances
TC 64	Electrical installation and protection against electric shock
TC 78	Live working
TC 81	Lightning protection
TC 99	System engineering and erection of electrical power installations in systems with nominal voltages above 1 kV AC and 1.5 kV DC, particularly concerning safety aspects
TC 122	Low-voltage switchgear and control gear

National Standardization and Certification (IRAM)

The Instituto Argentina de Normalización (IRAM) was founded in 1935. It is a private organization, which was recognized by the government in 1994 as an institute for standards in Argentina (National Decree 1474/94). As the national standardization organization, IRAM produces standards by itself or in conjunction with other organizations.

International standards from ISO and IEC are adopted by IRAM for Argentina whenever possible. Instituto Argentina de Normalización also subscribed to the 'Code of Good Practice' for the development of standards, defined in Technical Barriers to Trade Agreements of the World Trade Organization (WTO) (https://www.wto.org/english/docs_e/legal_e/17-tbt_e. htm, accessed 2 March 2016).

Instituto Argentina de Normalización is the first body accredited by Argentina's Accreditation Organization (OAA) for the certification of persons working with nondestructive tests (IRAM-ISO/IEC 17024, IRAM-ISO 9712). The Institute is a member of the IEC CB Scheme and IQ Net.

Instituto Argentina de Normalización has developed several international agreements on cooperation in standardization and certification. International agreements have been made with Brazil, United States, Venezuela, Spain, France, Ukraine and Malaysia. Agreements have been made with certification bodies: AENOR in Spain, ANCE in Mexico, ASTA in the United Kingdom, CEBEC in Belgium, CSA in Canada, ITS in the United States, JET in Japan, KEMA in the Netherlands, KTC in Korea, LCIE in France and TÜV in Germany, to mention some.

1.3.2.2.2 Chile
International and national standardization in Chile is relative young. Founded in 1973, the Chilean National Standardization Institute (INN) started as a private entity. It represents Chile in ISO today and is preparing its membership in IEC. The INN is part of the national quality system and has activities in standardization, accreditation, metrology and education.

On the government side, standardization in Chile is linked to the Ministry of the Economy and the Ministry of Energy. These Ministries have formed the National Commission of Energy

(CNE), the Commission for the National Energy Efficiency Program (PPEE) and National Governmental Supervisory Body (SEC) to support the government in Chile on questions of standardization and certification.

The National Standards Institute (INN) is the government-recognized organization for standardization and certification. It has established technical committees with members for customers, private entities, universities and laboratories to work on standards. Through the Ministry of Economy the work of INN may also be published as a regulation (www.inn.cl, accessed 7 February 2016).

The Association of Electrotechnical International Industry (EPEI) (Associación Gremial de Empresas Internacionales de Productos Eléctricos) is the National Committee representing the country on the IEC. The EPEI members are Osram, Siemens, ABB, Schneider, Legrand, BTICINO, and Philips.

1.3.2.3 Andean States

The Andean states are Venezuela, Colombia, Ecuador and Peru. In standardization activities Colombia has the lead. The participation of Andean states in standardization on an international level is relatively low and in most cases international standards are applied.

1.3.2.3.1 Venezuela
The standardization organization in Venezuela was Fondo Norma which was replaced by the National Standardization Body (FDN). The FDN carries out activities in fields such as safety and occupational health and quality management. In the electrical field, standards are released in four technical areas as shown in Table 1.10.

The FDN of Venezuela has only a few standardization activities of its own. Most standards are used from international organizations like IEC and ISO and standards from the United States – for example, the IEEE.

1.3.2.3.2 Colombia
Standardization in Colombia is organized by the Instituto Colombiano de Normas Técnicas (ICONTEC). This is the government-recognized national standardization body representing Colombia in ISO, COPANT, PASC, IQNET, on the management board of ASTM and as an observation member of IEC. Under the Ministry of Commerce ICONTECT carries out two responsibilities: national standardization and organization of accreditation.

National accreditation uses the Deutschen Akkreditierungs-Rat (DAR) for the Colombian quality management system. Accreditation agreements exist with Chile (WN) and Peru (INDECOPI).

Table 1.10 Technical activities in the electrotechnical field.

Committee	Title
SC 02	Electronics
SC 07	Electric installation
SC 08	Protection equipment
SC 09	Machines and parts

International agreements for cooperation in standardization have been made with Sweden, Spain, Argentina, Mexico, Norway, Brazil, the United States, Uruguay, Venezuela, Ecuador and Germany.

1.3.2.3.3 Peru

National standardization in Peru is carried out by the National Normalization Institute (INDECOPI). Peru follows international standards such as ISO and IEC and standards from ANSI, IEEE and NEMA in the United States are widely applied.

There are some national standardization activities for electric motors, illumination, cables, low-voltage switchgear and power transformers. The healthcare sector is regulated by the Health Ministry (Ministerio de Salud).

The National Institute of Defense of Competition and Protection of Intellectual Property (INDECOPI) is the national standardization body of Peru, which has a suborganization – the Peruvian Office of Normalization (OPN). The OPN operates some technical committees for normalization. There are three categories of standards and guidelines used in Peru:

- Category A: international technical standards, for example ISO, IEC, ITU. These standards are applied directly in Peru.
- Category B: international technical standards in the final phase of approval, for example draft international standards. These are used as draft standards in Peru.
- Category C: regional technical standards, such as CEN, CENELEC, ETSI, COPANT, ANDINA. These are used directly in Peru.

1.3.3   Asia and Oceania

1.3.3.1   General

Standardization in Asia and Oceania is very different in activities and organizational structure. One of the first Asian nations to use standardization is Japan, which introduced it in conjunction with its industrial development. Today China is growing in national and international standardization activities, with an increasing number of Chinese experts attending international standardization organizations. India is an upcoming internationally active nation in standardization. With its links to Britain arising from the years of colonialism, economic growth and its innovative technology sector will drive India more into standardization.

In Australia the standardization structure is at the same level as in Europe or North America. Compared to the size of the country, its presence in international standardization activities is high. Other Asian and Oceanian countries are more followers of international standardization with few activities of their own.

1.3.3.2   China

The structure of the Chinese legal system is the basis for standardization activities in China. The principles are shown in Table 1.11.

The National Parliament of China is the highest legal authority and its laws have an impact on standardization. With regulations released by the State Council, more administrative

Table 1.11 Chinese legal structure.

Legislative	Documents	Type
National Parliament of China	Laws	Law
State Council	Administration	Regulations
Peoples' Congress or Standing Committees of the Provinces, Regions, or Municipalities	Local regulations	Regulations
Ministries and Commissions	Department rules	Rules
Standardization Administration of China	Mandatory standards	GB or DL standards

regulations are influencing the technical standardization process. In principle this is the same in Europe and in North America but the Chinese government is stricter in formulating requirements into laws.

The Peoples' Congress, which meets in the Great Hall of the People, and its standing committees have a strong impact on local regulations applied in Chinese provinces, autonomous regions, and municipalities directly under the central government. These regulations have a regional impact on standardization.

The State Audit Administration and other local organizations and authorities set up rules that impact standardization. These rules are set up at ministry level or by commissions of the State Bank of China.

Standardization and conformity assessment in China are based on several laws and regulations. Standardization laws were released in 1988 and 2007 and regulations were implemented in 1990. These laws and regulations are the basis for more specific regulations in the industry sector, at the national level and at local level.

On the conformity assessment side, a product quality law from 1993, revised and extended in 2010, and the Import and Export Commodity Inspection Law from 1999, revised in 2002, form the basis for more specific and detailed regulations regarding product quality certification from 1991 and the Implementation Regulation from 1992. The China Compulsory Certification (CCC) regulations have been published since 2001 for different areas of application.

Several laws and regulations have been formulated for environmental protection, as shown in Box 1.3

Legislation in China involves a permanent process producing new laws and amendments to existing laws for compulsory product certification (CCC). The liability of manufacturers for their products is based on preventive principles. The product liability laws require strict product liability for the producer; in the case of a fault there is fault-based tort liability. In a contractual relationship, there is 'contractual liability'.

The following requirements are connected to strict liability:

1. There must be a product. (Buildings and agricultural structures are not products.)
2. There must be a failure of the product.
3. There must be damage to a person or property.
4. There must be causality between fault and damage.
5. There should have been no relief to the author for the damage.

Box 1.3   Laws and regulations on environmental protection.

Laws
Energy Conversion Law
Renewable Energy Law
Energy Law
Electricity Law
Product Quality Law
Clean Production and Promotion Law

Regulations
Bylaw on Accreditation and Certification
Regulation of Energy Efficiency Labelling Administration
Decision on Energy Conversation by the National Council
Implementation Rule on Energy Label of Refrigerator
Implementation Rule on Energy Label of Air Conditioner
Implementation Rule on Energy Label of Washing Machine
Implementation Rule on Energy Label of Unitary Air Conditioner
Implementation Rule on Energy Label of Motor
Implementation Rule on Energy Label of Gas Water Heater

Figure 1.4   Responsibility for standardization.

Responsibility for Standardization

Overall responsibility for standardization in China rests with the Administration Authority of China (SAC), as shown in Figure 1.4.

The Standardization Departments of Ministries or Commissions release national and branch standards. Regional standards are published by standardization departments of provincial governments. Independent enterprises' standardization organizations release enterprise standards.

In China today there are more than 320 standard institutes and 290 national standards technical committees publishing standards. In addition to this, the China National Institute for Standardization

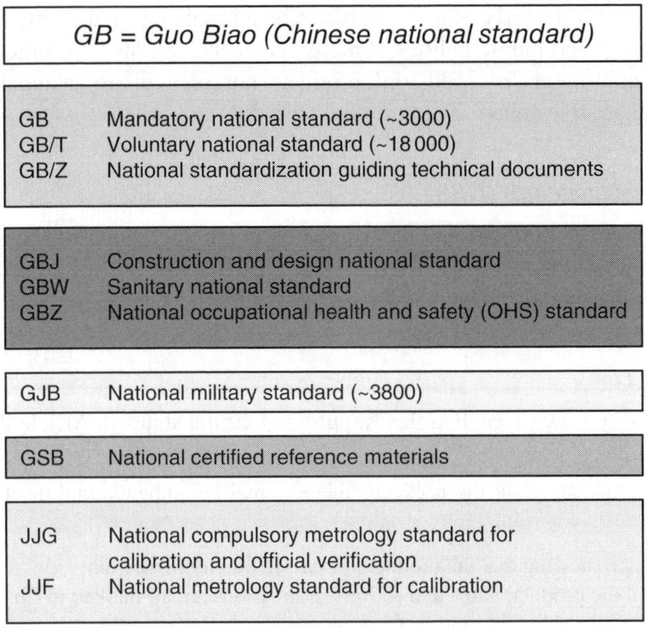

> **GB = Guo Biao (Chinese national standard)**

GB	Mandatory national standard (~3000)
GB/T	Voluntary national standard (~18 000)
GB/Z	National standardization guiding technical documents

GBJ	Construction and design national standard
GBW	Sanitary national standard
GBZ	National occupational health and safety (OHS) standard

GJB	National military standard (~3800)

GSB	National certified reference materials

JJG	National compulsory metrology standard for calibration and official verification
JJF	National metrology standard for calibration

Figure 1.5   Types of national standards.

(CNIS), the China Association for Standardization (CAS), the China Standards Information Centre (CSIC) and China Standards Press (CSP) are also active in the standardization process.

In the field of branch standardization in China there are 26 research institutes and specialized ministerial and commission standardization institutes; 339 technical committees publish documents and 12 specialized ministerial standards associations are active.

There are 158 provincial standard-publishing research institutes with an unknown number of regional technical committees. In parallel with these, 85 provincial associations are active in standardization. The number of enterprises active in standardization and publishing enterprise standards is not known. There are many enterprise standards from technical research institutes, technical committees and business enterprise associations.

China-wide, there are about 38 000 members of standardization technical committees; 100 000 members of research institutes and universities work together with more than 20 000 members of the standardization management standing organizations. They work on more than 3000 mandatory and 18 000 voluntary standards at the National Standards Level Guo Biao (GB) – see Figure 1.5, which shows the different types of national standards (GB).

Mandatory Standards

Mandatory standards are fixed in Article 7 of the Standardization Law of the Peoples Republic of China. National standards and trade standards are classified into compulsory standards and voluntary standards. Those for safeguarding human health and ensuring the safety of people and property and those for compulsory execution as prescribed by the laws and administrative rules and regulations are compulsory standards; the others are voluntary standards.

Local standards formulated by the standardization administration departments of provinces, autonomous regions and municipalities, directly under the central government, for the safety and sanitary requirements of industrial products are compulsory standards within their respective administrative areas.

Origin of National Standards (GB)

There are more than 25 000 national GB standards in China. About 50% have been adapted from international standards, mostly from ISO and IEC. The rest of the adapted GB standards come from other organizations such as DIN/VDE or IEEE.

Product Quality Law

The Product Quality Law of the Peoples Republic of China states in Article 27:

Marks on the products or on the packages thereof shall be authentic and meet the following requirements:

1. with certificate showing that the product has passed quality inspection;
2. with name of the product, name and address of the producer, all marked in Chinese;
3. with corresponding indications in Chinese regarding the specifications and grade of the product, and the name and quantity of main ingredients, where such particulars are to be indicated according to the special nature and the instructions for use of the product; with corresponding indications on the package of information necessary for consumers to know in advance, or providing consumers in advance with documents indicating these information;
4. with production date, safe-use period or date of expiry at easily spotted areas if the product is to be used within a time limit;
5. with warning marks or statements in Chinese for products that, if improperly used, may cause damage to themselves, or may endanger the safety of human life or property.

China Compulsory Certification (CCC) is organized into different fields of application as shown in Box 1.4.

These accredited certification bodies use more than 20 different marks for certification indications as shown in Box 1.5.

Box 1.4   China Compulsory Certification (CCC) – Accredited Certification Bodies.

CCC	Basic mark
CCC	Product safety
CCC	Fire protection products
CCC	Product safety and EMC
CCC	EMC
CCC	WLAN products

Box 1.5   Certification marks used by China Quality Certification Centre.

Safety Product Certification
General Product Certification
Safety and EMC Product Certification
EMC Product Certification
Performance Product Certification
ISO 14000 EMS Certification
Energy Save…Product Certification
Textile Safety Product Certification
Bolts Certification
Environmental Label Product Certification
Hazard Analysis and Critical Control Product
IECEE CB Scheme
IQNET
Organic Product Certification

1.3.4   Africa

Standardization activities in Africa follow the international standards of IEC, ISO or ITU and regional standards from North America (e.g. IEEE or ASME) or from Europe (the EN standards). In recent years standards from China have appeared in Africa. In Africa, only the Republic of South Africa is active in international standardization work related to industrial development there.

In northern Africa, Egypt plays a role in standardization with participation in international standardization.

National standardization in African nations follows the international standards in most cases. Driven by international consulting companies and the World Trade Organization, with financing from the World Bank, international standards of the IEC, SO and ITU are used primarily.

1.4   The Public View of Standardization

The general public view of standardization is rather negative. As soon as someone states: 'I'm active in standardization!' people will think someone who is rather dull and grey. It gets a little better when one states: 'I'm active in international standardization!'. At least – thinks the general public – he can travel to some exotic places like China, Japan, India and America. But in both cases the general public thinks that his professional career is coming to an end in his company's hierarchy.

One little piece of the pattern that forms such pictures in the mind of the general public mind relates to standardization activities. For example, the front page of a standard – which is also standardized and needs to be standardized for easy recognition by users – sometimes gives the impression that standardization experts sit around all day with a ruler, measuring millimetres and centimetres on the title page of a standard as Figure 1.6, from a German standard, shows.

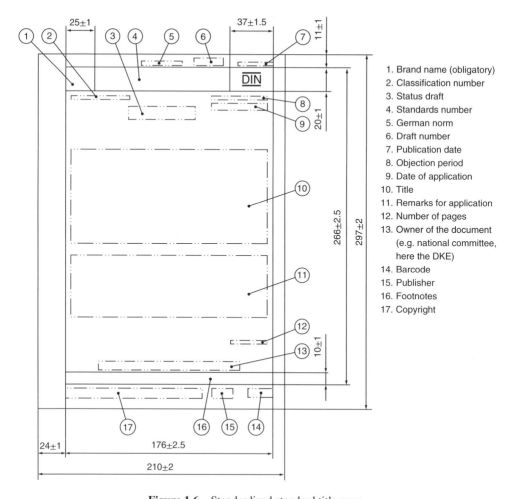

Figure 1.6   Standardized standard title page.

The information given on the title page is well structured and there is a need to simplify the world of technical information by using structures.

The general public also pictures an expert at one of the standardization institutes as a white-collar worker, with a very strict appearance. Such work is seen as bureaucratic and not really interesting or exciting. This picture is still too often the public view of standardization.

Standards are also seen as somehow trivial – for example, European requirements about the size of apples or the radius of bananas or cucumbers. Such examples can be found all around the world.

What the general public does not know is the complexity of industrial standards, the high grade of innovative technology related to the latest versions of the standards, or the new technical fields made accessible to the market. This is made possible by experts in the related technical field around the world. The people engaged in standardization work are the best in their technical fields.

The general public's knowledge of standardization is not very specific – people do not know how it works – but many would agree on the statement that standards benefit them. For most people standards are seen as a basis for economic success. Products and services are easier to provide with standards than without. Particularly in Asia, standardization is seen by the public as a way to deliver to a global market.

In Japan, standardization is more important to the public than it is in the Western world. This high level of interest even led Japanese TV makers to produce a documentary on standardization that was broadcast at 8 p.m., prime time. It showed that Japanese industry is producing products for the world market and related how there was a need for the country to be active in international standardization, to put forward its specific interests and technical views. The show illustrated how Japanese engineers are travelling around the world to participate in experts' meetings, providing information that would influence the content of standards.

On the other hand, the topic of standardization is often missing from engineering education in universities. This should be improved given the importance of standards in industry.

1.5   Right Timing

1.5.1   General

In standardization, if experts meet too early in the product innovation process they might find that not enough technical information is available on the product to specify requirements and to define test routines. On the other hand, if the experts are too late in starting work on standardization, various technical solutions might already be established in some market places and a unique standard solution might not be possible any more. The correct timing of standardization is related to the innovation process, market access and the regulatory processes as explained in the following.

1.5.2   Innovation Process

The innovation process is the driving force of technology companies. Here, new ideas are born that will bring new products onto the market, or improve existing products, or reduce prices. In some cases the innovation process brings all three in one.

In the beginning there is an idea about how to solve a problem, how to increase functionality, or how to avoid cost. These ideas are usually protected by patent rights. Innovative companies are proud of the number of patents that they receive every year and patents are among the goals of management. The belief is that only by producing enough patents each year can technology companies match their competitors in the market. But patents by themselves are not sufficient to be successful in the market. Standards are needed for this.

Many patents are needed to create one successful product in the market and it is not clear which of the patents will make it to a successful market product. To link these processes it is necessary to have a systematic view of the ideas and patents together with the design and development process. Within the product design and development process technical milestones are set to plan market access. These technical milestones provide orientation and a starting point for the standardization process. With standardization times of about 2 to 3 years, orientation in the design and development process comes with product prototypes and pilot installations. By this time enough technical information is available to explain technical requirements for the product and devise procedures to test these requirements. The standard writing process then can take place in parallel with the product marketing initiatives and the first drafts of the standard can be used for product specification of applications.

In international standardization, consensus standards are the goal and more than one manufacturer is needed to find this technical consensus. In most standardization processes rules are given on how many countries, industry sectors or users need to send experts to the new working groups to produce a standard. In the IEC, a minimum of four or five experts from different countries are required to start work, depending on the number of member countries in the related technical field. These rules are applied in any consensus-orientated standardization and give a balanced set of standard-writing expert groups.

So, besides the right timing, it is also necessary to find other manufacturers to be willing to participate in the standardization work. Having more than one manufacturer also increases the transparency of the standard's technical content. This allows any user to have a choice of different technical solutions and encourages price competition. The standard provides the basis for long-term success in the market.

Monopolistic markets, with one product and one manufacturer, do not need standards. Such situations happen in markets like smart phones and computers. Innovative industries provide different technical solutions and offer a wide range of products based on the same standard. The role of standardization is to set common rules for product requirements and test procedures. For this reason standards are not static and adapt with the innovation of new products. Depending on the industry sector, the mean time for the revision of standards is in the range of 3 to 6 years. For basic and horizontal standards the revision time may be as long as 12 years. Basic standards cover subjects that are very stable and do not change much. A horizontal standard is defined by the IEC as a standard that gives values and requirements for safety to a wide range of product standards – for example the levels of test voltages or the permitted electric and magnetic field strength to which humans may be exposed.

Innovation is the driving force for the future success of technical companies and consensus standards prepare and open the market for business.

1.5.3   Market Access

There are different ways to access the market with products. In a regulated market products come to market only if laws and regulations are fulfilled. In an open market, competition between manufacturers of technical products offers a wide range of products for customers' needs. In most cases market access involves a mixture of regulatory requirements and competitive products. What is the role here of standards in market access?

Standards are made to assist users to find the right products for their requirements and to enable them to trust that this product will have the required technical features.

An example from my own technical field, high-voltage switchgear, explains this. The electric power supply needs substations with switches to switch electricity on and off. In the high-voltage transmission network, such switches need to switch 420000V in Europe or 550000V in America and Asia and up to 1 100000V in China. They need to be able to interrupt rated currents of 3000A, 4000A and sometimes 5000A. In the event of a short circuit in the network, they are required to interrupt 50000A, 60000A or even 80000A. Worldwide there are some 20–30 manufacturers who produce high-voltage circuit breakers. Users of such applications are usually not in a position to prove the statements of manufacturers that their product is able to fulfil the criteria.

Standards now define the requirements and set rules for design and construction. In specific tests, which are usually carried out by independent test laboratories, standardized test processes are applied to prove functionality.

A positive test result is then documented in a test report and this test report is then used with the offering of a circuit breaker by the manufacturer. This give the user confidence that the product offered will function as required.

The standard is opening the market for products that follow the specified requirements. Without such a technical standard it would be very difficult for a manufacturer to convince the user about the functionality of his product. To carry out an approval test with each user in each project would be very expensive and this would be the situation without standards.

1.5.4   Regulatory Processes

In many countries a regulatory process sets additional requirements for products in cases where safety is involved. This could be electric shock, explosions, fire and other dangerous product effects to humans. Regulations and laws must be followed in every case; the application of standards is for the user to decide.

When international standards are written they usually follow any relevant requirements in so far as such requirements are known and acceptable to the expert group. On the other hand, regulatory bodies follow the recommendations of the standardization experts with their regulations and laws. Therefore, in most cases standards are in line with regulations. It is clear that there are technical areas where there is a difference in the requirements of regulations and standards. This conflict will stay because laws and regulations are produced by governments and their associated organizations, which make their decisions independently from the standardization organizations and their experts.

But standards are always oriented towards the regulatory requirements which are usually focused on human safety, environmental protection and health. Regulations, when made in a proper way, leave the technical specification to the standards and give only the framework for standardization.

1.6   Benefits of Standards

What are the benefits of standardization? This is an often-asked question in companies when it is necessary to justify why so many experts need to go to standardization working groups at distant locations for such a long time and so often. International standardization is often viewed in terms of long-distance travelling to such pleasant places as New York, Berlin, Paris, London, Tokyo, Delhi, Beijing, Rio de Janeiro…the list can be extended to many more.

In my company, in the field of energy transmission and distribution, more than 350 experts are active in standardization. As coordinator, responsible for the participation of our experts in the right standardization working groups, I often have to answer questions like: 'Is it necessary that these experts are participating in standards?' 'What is the benefit for our company?' 'What could be stopped without having a negative impact on our business?'

These questions are not easy to answer because the direct impact of standards can hardly be seen. Their impact is usually felt over some months or even years but the costs of travelling and working time have to be paid right now.

The interesting thing with standardization work is that even when all experts do a good job, when everything is running smoothly, when newly designed and developed devices can be sold on the market with reference to international, regional or national standards and when the business runs well – the experts in standardization are still asked what they are doing, besides travelling and whether it is necessary to send so many experts to so many standardization meetings. Again the benefits of standardization are questioned.

Unfortunately, experts in standardization working groups are immediately asked about their standardization work if equipment or a device that has just been developed does not fulfil some of the requirements and high additional development cost are necessary to bring the device or equipment to a technical level to fulfil the standard's requirements. Standardization experts can become very famous if something goes wrong but if everything goes well only the travelling is recognized.

A general understanding of the importance of standardization is seen in almost any company. Most are just following technical requirements as written in standards – they are passive users and do not participate actively. Companies that will become leaders in their technical field cannot act as passive user of standards; they need to be active creators of standards so that their leading technology is reflected in the standards. For them the main benefit is to have a standard in place at the time when the new designed and developed device or equipment is ready for the market. Market entry in parallel with new or revised standards creates a business advantage over other market players who cannot fulfil the new standards' requirements. This is the reason why standards are constantly adapted to technical developments. In this way, standardization work is more a strategic tool for successful business planning, understood as an important link between innovation and product design and acceptance of new product in the market when based on standards. Technical leaders always work that way. Beside this strategic view of standardization, the general benefit of standards can be described, as reported in an investigation by the Deutsches Institut für Normung (DIN) in 2010 [2].

In principle, there are two areas of benefit: business benefits and social benefits. Business benefits cover the strategic benefits explained above – advantages in a competitive market (national, regional or international), reduction of product costs, reduction of costs of relationships with subsuppliers, an increase in safety and reduced costs for product liability, a link from research to market and an opportunity to make technical developments known to the public.

The benefit of standardization for society is seen in the political sphere, national economics, international trade and macroeconomic advantages for companies active in national or international business.

The business benefits were investigated in Germany, Austria and Switzerland [2]. For this, more than 4000 companies, organizations and institutions were interviewed and the results were presented. At the strategic standardization level it was found that the marketing effect is seen by standardization experts but is often not recognized by the companies' management. This result is that standardization activities are recognized as expenses ('they are travelling all the time!').

Seventy-five per cent of those questioned indicated that the experts are engaged in international (ISO, IEC, ITU) and regional (CEN, CENELEC, ETSI) standardization to influence standards work.

Standardization activities need to be planned on a long-range scale to create benefit for the companies. Most short term engagements are not effective and cannot create benefits. The main advantage seen by the companies is that they can offer products at the time of the publication of the standard whereas competitors not active in standardization need to redesign their product and come to market with a delay and a higher development cost. This effect is even stronger if the regulator uses standards for defining regulations, which is quite often the case.

Direct advantages of companies in the technical field over their competitors are not often seen with consensus standards as they are usually also participating in the standards process. So active participation in standardization is more to avoid a competitive disadvantage that would arise if a product is not in line with a standard. Standards set by companies (IBM, Apple, Microsoft) can create a strong advantage in competition against other companies in the market. They can also give rise to disadvantages in creating a market as customers do not like monopolistic situations and it is difficult to reach such a market position. In many cases such company standards do not last for a long time.

The main advantage of standardization seen by companies is the time and knowledge advantage. Those who participate in standards are informed at an early stage about the changing technical requirements in a standard and they learn the reasons why technical changes are made from their competitors participating in the standardization work.

The influence of a business's own experts participating active in standardization is seen as high. Those who write the standard's text first have the highest impact on the final text. In Germany we say: 'Wer schreibt, der bleibt' ('who writes the text, stays with the text').

Global business requires global standards. Even small and medium-size companies are using international markets for their products to sell or to buy subparts for their own use on the global market. The Internet offers global sourcing for anybody. The standards offer the required quality level and reliable products. Small companies do not need to travel to Australia, the United States or China to test a product. They only need to use globally accepted standards to sell and buy products. That is why the main focus on standardization is at the international level with a global reach. The costs of adapting products to regional or national standards are usually high. Redesign, testing and certification can easily cost several hundred thousand euro or more, which must be earned back in the regional market.

In a harmonized standardization field many additional costs can be saved and cross-border trade is much easier. Standards open up many trade opportunities around the world with easier contracts and fewer trade barriers.

To compare cost for experts active in standardization with costs arising from failing to fulfil standard requirements and necessary redesign is not easy because companies do not like to speak about that, or publish numbers.

When a company misses a standard requirement and is unable to bring a product to market successfully, this will not be communicated to the outside world. Experts will sit down and will change their product so that it is in line with the standard or they will stop selling the product.

The problem with sending experts to standardization work is that these experts are also needed in the company for the development of the new devices. This is a standards expert's dilemma all the time.

Cost reductions in companies through the use of consensus standards are usually not appreciated. Standards are often just seen as requirements to be fulfilled. The cost-cutting effects are also not communicated with competitors because they give businesses advantages in competitive markets.

In the subsupplier market, there is much less dependence on one supplier in standardized technical fields than in fields without standards. There are more suppliers and there is more competition. The standards require quality and reliability, which has to be delivered by the subsupplier and this is an advantage in the market. The supplier can share a much larger market and he does not really need to know the final use of the delivered product. The market power of the standardization of products is much stronger for the supplier market than for the user market.

Standardization supports cooperation between companies as technical requirements are defined and quality controlled by testing, giving reliable products.

From a research and development point of view standardization does not affect the innovation process because it comes later in the process chain. Standardization of research work does not really make sense because there is not yet a product or service for the market. Standardization starts with the market process of developing products and services ready to be used in the market.

Participation in the standardization process during product development reduces the risk of stranded investments because the content of standards can be influenced by discussions with the other experts in the standardization working group. A company's own misdevelopments can be detected early and can be corrected or stopped.

Typical reaction times for new developments and adaptation of standards are in the range of 2–5 years. Two years' revision in consensus standards is the shortest possible time when following the commenting and voting rules. In most cases 3–4 years are needed.

In terms of insurance, standards are mainly seen as safety. In companies using standards, injuries and accidents are much lower than in companies not following standards. Safety-related requirements are covered by standards for all types of products and help to reduce the liability for nonfunctioning products.

The economic benefit of standards is not only linked to the innovations made in society – it also needs a competitive market with efficient produced and reliable products. The economic benefit is created by the efficient diffusion of innovation and standardization.

Investigations in Germany have shown that in very innovative technical areas (in terms of number of patents per year) a high number of standards are also produced. The dynamic development of innovations creates new standards and the faster the innovations occur the quicker new standards are developed. The standardization process reflects very much the type of business behind it. Very dynamic companies with a high level of innovation are often not very open to participating in consensus standardization because they fear to loose knowledge to competitors.

In a long-term comparison of the number of patents and standards from 1960 to 1996 it can be seen that standards influence technical innovation and market development just as patents do.

Exports to the global market are easier with international standards. Standards do not only affect exports; they influence imports in the same way. So there is a winning situation on both sides for the user. Exports based on standards increase; imports based on standards increase.

Standards are very modern and up-to-date. The following subsections give some examples of today's standardization activities with benefits for companies and society [3].

ISO/IEC 15962 Radio Frequency Identification (RFID)

With this standard, automated identification and localization of things or livestock can be made. It is easy to find objects in a factory or storage, or to control the feeding of animals, or to identify flowers from growth to your table at home. Radio frequency identification technology allows IT support for the logistics of any object – the flower from the Netherlands or South Africa, or the seat of an automobile from the subsuppliers to car assembly.

IEC 62478 Smart Textiles

This standard defines safety requirements for smart textiles related to the person wearing such clothes. Textiles with specific functions are possible in work or in private activities. Control of body temperature, electromagnetic fields, changing colours, energy generation and heat protection are only a few possibilities for new business.

EN 60601 Telemedicine

Telemedicine can be used by a doctor far away can collect data about your health or give information about medicine. It can be used to support elderly people in controlling their own lives or to assist with complicated surgery by consulting a doctor far away. This standard gives requirements and quality-control measures for telemedicine.

DIN VDE 0100-705 Big Data

Collection, management and evaluation of large amounts of data is becoming much easier with data storage possibilities available today. Big data include weather data, data regarding livestock management including the control of each cow and anything it will eat and data from sensors on human bodies collecting data about any move. This standard defines data handling and privacy of data.

IEC 61850 Smart Energy

The Internet of energy needs to be managed over a wide area with millions of devices generating, transmitting and distributing electrical energy with guaranteed network stability and efficiency each millisecond. This standard is an information backbone for the future renewable electric power supply.

EN 61508 Smart Workflow

This standard is the basis for equipment such as a robot in an assembly line of a car manufacturer, which poses a risk of causing harm to workers.

EN 61950 E-Mobility

The e-mobility of cars, bikes and bicycles with an electric battery storage and notes. Charging station and vehicle safety. The basic aspects are defined including standardized plugs and charging processes.

IEC 62883 Ambient Assisted Living (AAL)

Using modern technology to help older people to manage their daily lives in their own homes. Electric motors to control window shades, automated vacuum cleaners, service robots, stair mover, alarm in case of emergency, motion detection and many more technical aids are defined in this standard.

ISO/IEC 24571-1 Smart Collaboration

This standard defines requirements for e-learning software, e-books, online studies with examinations and virtual lectures.

2

Framework for Standards

2.1 General

This chapter explains the global framework for standardization. It considers the different effects of governments and national political and industrial organizations. In particular, it considers the World Trade Organization (WTO), the European Union (EU) and regional structures in the United States, China and Europe.

2.2 World Trade Organization

2.2.1 General

The World Trade Organization (WTO) is an international organization with its headquarter in Geneva, Switzerland. The WTO works to improve trade and economic relationships between nations. It was founded in 1994 as a logical consequence of the General Agreement on Tariffs and Trade (GATT) after a 7-year consideration in the so-called Uruguay Round.

Beside the International Monetary Fund (IMF) and the World Bank, the WTO plays an important role as a centralized international organization. It addresses economic matters that have a global reach.

As a matter of principle, the WTO does not become involved in national product and service regulations. This is left to the national governments. In terms of standardization the WTO does not allow national standards, which serve as international trade barriers. Removing these national trade barriers and encouraging the acceptance of international standards by national laws and governments is the main role of the WTO. The WTO prefers situations in which national governments set up laws that require the use of international standards. However, this would be too strict and would hinder the ongoing technical development of products and

Practical Guide to International Standardization for Electrical Engineers: Impact on Smart Grid and e-Mobility Markets, First Edition. Hermann J. Koch.
© 2016 John Wiley & Sons, Ltd. Published 2016 by John Wiley & Sons, Ltd.

solutions. The practice today is to leave it open to the manufacturer to decide whether to apply standards; this gives more freedom for product specifications and does not limit it to what is written in standards. The standardization process for new standards or for revisions is too slow to follow technical developments and laws are even slower. Typical standardization cycle times are in the range of 3 to 5 years.

Nations are encouraged by the WTO to be active participants in international standardization and to contribute their competences and experiences. This is important if there is to be globally accepted standardization.

National standardization organizations, in the view of WTO, should reach a national consensus on the matter treated in an international standard. This principle takes into account that each nation has technical, economic and social organizations that are able to collect opinions nationally and find a consensus. National consensus is a basic rule for an accepted and applied standardization.

The WTO also uses this approach to coordinate the consistency of standardization work at international and global levels. Standardization work is made public and a public inquiry process allows comments to be made within a 2-month period. In this way the WTO contributes to international standardization with the aim of improving trade between nations to the benefit of the people.

The United Nations has three economic commissions (UN ECs), which promote the economic integration of member nations and coordinate their work with other regions of the world. These commissions cover the following topics: economic cooperation, trade, statistics, sustainable energy, technical cooperation, environmental policy and transport. Each topic has its working groups – the important one for standardization is the trade section. This deals with regulatory cooperation and standardization policies. Topics are identified where governments see the need for a transnational dialogue on a regulatory convergence matter. Standardization and regulatory activities are presented in a list and kept up to date.

The following five economic commissions exist in the United Nations:

• United Nation Economic Commission for Europe (UN ECE);
• United Nation Economic Commission for Africa (UN ECA);
• United Nation Economic Commission for Latin America and the Caribbean (UN ECLAC);
• United Nation Economic and Social Commission for Western Asia (UN ESCWA);
• United Nation Economic Commission for Asia and the Pacific (UN ESCAP).

2.2.2   United Nations Economic Commission for Europe (UN ECE)

The United Nations Economic Commission for Europe (see Figure 2.1) was founded in 1947. A total of 56 nations are members. These are European countries, as expected, but also the United States, Canada, Israel and Central Asian republics. This can be explained by the strong economic connections between these nations.

The main objective of its activities is to promote a policy, financial and regulatory environment conducive to economic growth, knowledge-based development and greater competitiveness of countries and business.

The standardization activities are linked to Working Party 6 (WP6), which is a forum for dialogue among regulators and policy makers. The participants discuss a wide range of issues, including technical regulations, standardization, conformity assessment, metrology, market

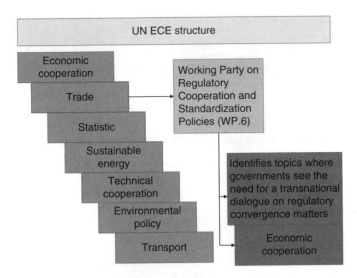

Figure 2.1   UN ECE structure.

surveillance and risk management (http://www.unece.org/trade/wp6/welcome.html, accessed 12 February 2016). Up to 2014, a total of 16 recommendations had been published; four were related to standardization:

- Further Development in International Cooperation on Technical Harmonization and Standardization Policies;
- Coordination of Technical Regulation and Standardization;
- International Harmonization of Standards and Technical Regulations;
- Reference to Standards.

It is a goal of Working Party 6 to avoid national laws and regulations for products and services and to promote international standardization. This harmonization process will lead to greater coordination in different areas of industry. In the end, trade barriers will be reduced and the free trade of goods will be the norm.

2.2.3   United Nations Economic Commission for Africa (UN ECA)

The mission statement of UN ECA is: 'Africans must seek growth that is primarily anchored on their priorities and that is capable of delivering structural transformation.' Founded in 1958 the UN ECA has today 54 member states.

The goal is to promote the economic and social development of its member states, faster interregional integration and to promote international cooperation.

The UN ECA provides technical advice to African governments, intergovernmental organizations and institutes. It is structured according to focus topics, including one on innovation and technology as shown in Figure 2.2. The African Information Society Initiative assists African countries and regional economic communities in the formation, adoption and implementation of new technologies that will help the transformation process in Africa.

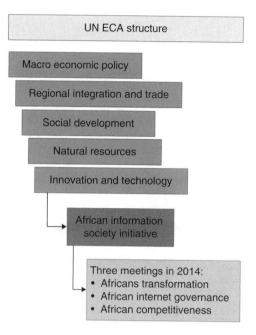

Figure 2.2   UN ECA structure.

The transfer of new and emerging technologies is supported by research initiatives from national and regional institutions. The main focus is on agriculture and social service delivery. Technical fields like electrical engineering have a lower level of importance. In 2014 the African Information Society Initiative held three meetings on African Transformation, African Internet Governance and African Competitiveness in Innovation and Technology.

The role of international standardization is not strongly connected to African countries. Only South Africa plays an important role in participating, with experts in those technical fields South Africa where has developed an industry – for example, electric power systems.

2.2.4   United Nations Economic Commission for Latin America and the Caribbean (UN ECLAC)

The United Nations Economic Commission for Latin America and the Caribbean was established in 1948 for South America. The Caribbean was added later. The Commission gas its headquarters in Santiago, Chile. The goal is to contribute to economic development in Latin America and the Caribbean.

There are 44 member states and 13 associated members. A session of all members is held every 2–3 years. The UN ECLAC has several divisions covering social and economic development, natural resources and infrastructure, production, productivity and management, and international trade and integration.

The task of the International Trade and Integration Division is to organize meetings on particular topics to develop the economic strength of Latin America and the Caribbean.

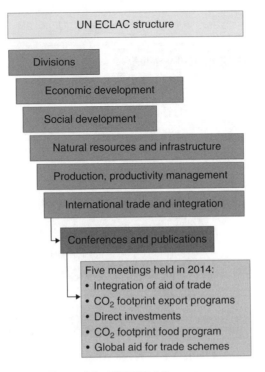

Figure 2.3   UN ECLAC structure.

The structure of UN ECLAC (Figure 2.3) shows that standardization does not play an important role. Countries are linked either to the IEEE or to IEC, depending on historical factors.

2.2.5 *United Nation Economic and Social Commission for Asia and the Pacific (UN ESCAP)*

The United National Economic and Social Commission for Asia and the Pacific (UN ESCAP) was established in 1947 in Shanghai, China. It represents 53 member states and nine associated members and covers an area from Turkey in the west to the Pacific island of Kiribati in the east, Russia in the north and New Zealand in the south. More than 4 billion people live in that part of the world.

The structure of UN ESCAP is divided into sections dealing with sustainable development, transport, social development, environment, information and communication, disaster risk reduction and trade and investment.

The trade and investment work is focused on foreign direct investments, business development, sustainable business practices, trade agreements, trade and investment research, trade facilitation and trade policy. See also Figure 2.4.

The Commission plays a central role in business development, economic growth and innovation. The Business Advisory Council (EBAC) gives technical assistance for economic

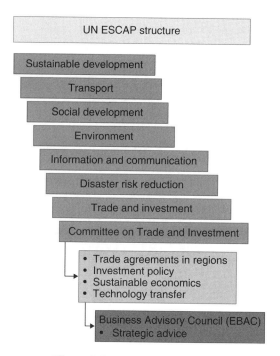

Figure 2.4   UN ESCAP structure.

developments in UN ESCAP. Founded in 2004 in Shanghai, China, EBAC provides business perspectives on development issues, active programs, and strategic direction.

The Commission covers an area with strong regional standard-writing countries like Japan, China, Korea and India. In this region the international standards like IEC, ISO, and IEEE compete with local requirements.

2.3   European Union

2.3.1   General

The European Union is a community of sovereign nations based on the EU contracts of all member nations. The seat of the EU parliament is in Strasbourg, France, and the EU Commission has its seat in Brussels, Belgium. The EU Parliament is elected by the member countries in a direct vote every 4 years. The EU Commission represents the governments of the EU member countries and executes EU Directives, which are transferred to national laws by the member governments on basis of the EU contracts.

The EU today has 28 member countries as shown in Box 2.1.

For standardization in Europe the Treaty of Nice has an important role in the voting on standards. The Treaty introduced double-majority voting. Double majority means that a vote passes if a weighted majority and the majority of the population have been reached. For this reason, each member country has a voting weight as shown in Table 2.1.

Box 2.1   Members of the EU.

Members:

Austria	Estonia	Italy	Romania
Belgium	Finland	Latvia	Slovakia
Bulgaria	France	Lithuania	Slovenia
Croatia	Germany	Luxembourg	Spain
Cyprus	Greece	Malta	Sweden
Czech	Hungary	Poland	The Netherlands
Denmark	Ireland	Portugal	United Kingdom

Proposed members:
Croatia
Turkey

Table 2.1   Weighted voting in the European Union.

EU members		Non-EU members	
France	29	Turkey	29
Germany	29	Switzerland	10
Italy	29	Norway	7
United Kingdom	29	Macedonia	4
Poland	27	Iceland	3
Spain	27		
Romania	14		
The Netherlands	13		
Belgium	12		
Czech	12		
Greece	12		
Hungary	12		
Portugal	12		
Austria	10		
Bulgaria	10		
Sweden	10		
Croatia	7		
Denmark	7		
Finland	7		
Ireland	7		
Lithuania	7		
Slovakia	7		
Cyprus	4		
Estonia	4		
Latvia	4		
Luxemburg	4		
Slovenia	4		
Malta	3		
Total EU vote:	352	*Total non-EU vote:*	53
Total CENELEC:	**405**		

This weighted voting system gives the big countries a higher weight. The six largest countries hold 170 votes. To reach a two-thirds majority, which is required to pass a standard in Europe, 255 votes are needed as stated in the EU contract of Nice. That means that 13 of the largest member nations need to approve the vote.

Experience with this rather complicated voting system within the European Committee for Electrotechnical Standardization (CENELEC) has been relatively positive. The largest countries in Europe are also the ones with the most industry and strongest participation in the standardization work. In the end, experts nominated to working groups by the national committees of CENELEC contribute with their personal knowledge and form the content of the standards documents. So a small country with only a few votes, like Denmark or Finland, can establish its technical view on the standard and will receive the majority of votes at the end. In most cases the votes result in very high approval levels – between 90 % and 100 % – because a technical consensus has been reached before the vote, in the working group.

2.3.2   European Commission

The European Commission represents the executive of the EU for EU directives. The EU Commission is the administrative body, led by the member countries. Each head of the Commission is proposed by the government of the EU member countries and must be approved by EU Parliament every four years after elections. The EU Parliament also needs to approve the EU budget.

The impact of the EU Commission on standardization is related to the EU Directives. The Directives give basic requirements for national laws and regulations of the member countries. It is mandatory that the governments of EU member countries apply EU Directive within one year of publication. In some cases it takes more time and for this a penalty system applies.

Standards released by the European Committee for Standardization (CEN) or CENELEC must apply to the requirements given by EU Directives. The overall goal of the EU Directives is to harmonize the EU market and the EN standards.

The conformity of goods or services brought into the EU market is indicated by the CE sign. The CE sign is a producer sign used by those who bring a product or service into the EU market. It is not a quality sign. Many EU directives have been published covering, for example, the shape and size of an apple, the dimensions of steps on a ladder, or the electromagnetic fields allowed for technical products. The EU Directives related to electrical standardization are shown in Table 2.2.

European Union Directives give principal requirements and EN standards specify these requirements with technical values and methods to test for product approval.

Table 2.2   EU directives in the electrical field.

EMC Directive (2004/108/EG)	Harmonizing electromagnetic compatibility requirements
Machinery Directive (98/37/EG) 2009	Harmonizing machinery requirements
Low Voltage Directive (2006/95/EG)	Harmonizing requirements of electrical equipment up to 1000 V
Pressure Vessel Directive (97/23/EG) 1997	Harmonizing requirements of pressure vessels

A principle in the European Union is that international standards from the IEC and ISO have a priority and must be taken into EN standards without any deviation if possible. There need to be reasons if a deviation is to be accepted by CEN or CENELEC.

European standards (EN) must be transferred into national standards with all members of CEN and CENELEC (see section 4.5).

Practical experiences of standardization work in CEN and CENELEC with the EU Commission Directives show that there is a close cooperation among the organizations. The harmonization of EN standards with the EU Directives is supervised by the New Legislative Framework (NLF), which uses a questionnaire process to decide if EN standards need to be revised or updated.

2.4 Regional

2.4.1 United States

Standards in the United States are seen as a common and repeated use of rules, conditions, guidelines or characteristics for products or related processes and production, methods and related management system practices. They can be split into voluntary consensus standards and nonconsensus standards, governmental standards or laws.

Voluntary consensus standards are written with openness, a balance of interest, with due process and allow an appeal process, to produce, at the end standards reached by consensus. The document is not published if consensus cannot be reached.

Nonconsensus standards are industry standards, company standards, consortia standards, or de facto standards. They have been developed by the private sector and do not follow the full consensus process.

Government standards are developed by the government for its own use, for example to set environmental standards, or standards are mandated by the law such as those contained in the United States Pharmacopeia and the National Formulary as referenced in 21 U.S.C. 351.

The actors and processes are shown in Figure 2.5.

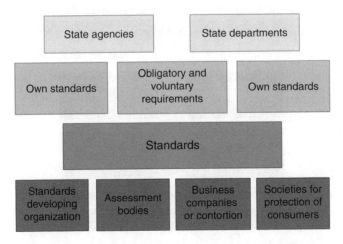

Figure 2.5 Actors and processes in the United States.

The actors in standardization are found at the governmental level in state agencies and state departments. They set their own standards and comply with obligatory or voluntary requirements related to standardization.

Industry and the consumer participate in the standardization through standard-developing organizations (there are more than 270 accredited organizations in the United States), assessment bodies, business companies or consortia and societies for the protection of consumers.

The standards developing organizations (SDO) in the United States cover a wide range of industries and are split into more than 270 traditional standard-developing organizations, which work on consensus standards under the control of American National Standards Institute (ANSI).

Today there are more than 10 000 consensus standards developed by NIST. More than 6500 have been developed by nongovernmental standard-developing organizations. In addition there are hundreds of 'nontraditional standard-development bodies' (consortia standards).

Consortia also reach consensus in their standards but only with those who are part of the consortia and not with other interested parties, as standard-developing organizations need to do to reach consensus.

The US standardization system is shown in Figure 2.6.

The National Institute of Standards and Trade was founded in 1901. It is a state federal agency within the US Department of Commerce (NIST). The American National Standards Institute (ANSI) was founded in 1918 as a subgroup for coordination and setting rules for the standardization process requirements. It is an administrator and coordinator for the US private sector in the field of voluntary standardization. It is the only accredited voluntary consensus standard-developing organization in the United States.

The American Society for Testing and Materials (ASTM) was founded in 1898. It has a database of 12 000 standards, which are used in 80 countries. The American Society of

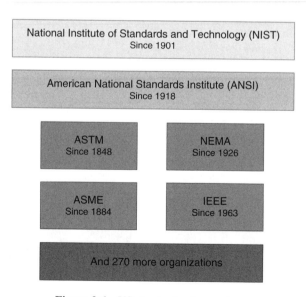

Figure 2.6 US standardization system.

Mechanical Engineers (ASME) was founded in 1884 and has more than 500 standards in its data base with annual reviewing. The prime goal of ASME is to support research work for future standardization.

The National Electrical Manufacturing Association (NEMA) is a federation with 430 member companies that manufacture products. It covers 50 sections of different products.

The Institute of Electrical and Electronic Engineers (IEEE) has more than 400 000 personal members and was founded in 1884. It has more than 1300 standards and is active in 160 countries through its members.

This little overview shows that standardization in the United States started with industrialization and therefore has a long history. The roots and content of standardization are both in America. Today, in international cooperation, ANSI is a member of IEC and ISO, organizing US participation in standardization work in the IEC and ISO as a national committee. The United States is one of the strongest members of the IEC and ISO with leading roles in IEC and ISO.

The standardization work organized by ANSI is decentralized through technical societies and standardization organizations, which cover different technical fields and are sometimes in direct competition. There are more than 600 societies and organization participating.

Beside these societies and organizations, there are 270 standard-developing organizations (SDOs), which have been accredited for writing standards on a consensus basis. All of these SDOs see themselves as international organizations because they also have activities outside the United States.

The federal structure of the United States does not lend itself to the coordination of these standardization activities inside the country. It is expected that the market will rule these organizations. This leads to different specifications in standards and makes it impossible to have unique testing and certification.

There is also no systematic transfer of IEC and ISO standards into US standards, as happens in the European Union by CEN and CENELEC.

2.4.2 China

The standardization system in China is based on laws released by the most powerful institute in China, the National Congress. A standardization law, statutes about certification and accreditation, statutes about product quality certification, laws about import and export inspection and many other laws and regulations have been put in place to control the standardization process, as shown in Box 2.2.

These laws, statutes and regulations are then used by the State Council, which is the government and administration, to set rules for the work of Ministries and their administrative departments for the standardization process in China. Since 2001 the State Council has also advised the General Administration of Quality, Supervision, Inspection and Quarantine in its work based on the same laws, statutes and regulations. The Ministry of Standardization sets up the Standardization Administration of China and the Agency of Certification and Accreditation plus 13 other administrative units not related to standardization.

In the provinces and local regions, additional administrative departments are active for standardization under the rules of the State Council. See Figure 2.7.

Box 2.2   Rules in China set by National Congress.

National Congress
Standardization Law
Statute about certification and accreditation
Statute about product quality certification
Law about import and export inspection
Other laws and regulations

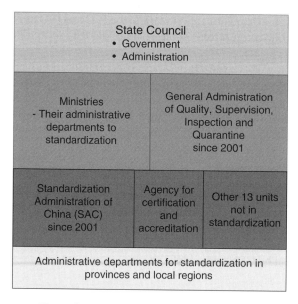

Figure 2.7   Standardization by State Council.

The Standardization Administration of China (SAC) manages the standardization process organized in regions, by industry and at the national level. The Administrative Department for Standardization in provinces and local regions therefore releases standards all over China, with contents that may vary in the regions according to specific requirements there – for example, climate, altitude, susceptibility to earthquakes, or other regional impacts. The Administrative Department for Standardization in Ministries and Industry releases standards related to the specific requirements of Ministries, for example health, mining, oil and gas, or for specific industry needs, such as for steel, coal, or semiconductors.

In parallel with these standardization activities the National Institute of Standardization (CNIS) releases standards of national interest, for example on electric power transmission or hydropower generation. The Standards Press of China (SPC) publishes the standards and since 2001 the China Association for Standards (CAS) is publishing standards as well.

At the international level (IEC or ISO), the Standards Administration of China (SAC) represents China as the Chinese National Technical Committee of Standardization. This complex structure is simplified in Figure 2.8.

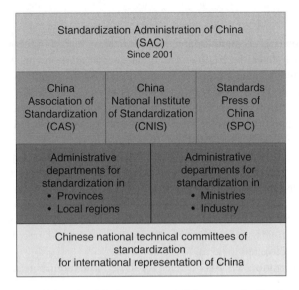

Figure 2.8   Simplified structure of Chinese standardization.

Figure 2.9   Types of standards in China.

Different types of standards are released in China. National Standards may be mandatory or voluntary in application and are published by the Standards Association of China (SAC). Trade or professional standards are released by competent administrative authorities under the State Council and may be mandatory or voluntary. The local departments of the Standards Association of China (SAC) release local standards, which may be mandatory only within local areas. Finally, there are Enterprise Standards, which are mandatory for enterprises or groups of enterprises. These different types of standards are found in different regions and in the various industries, see Figure 2.9.

2.4.3   Europe

The European Union (EU) is another large standardization region in the world. Consisting of independent national states the EU had one great goal in harmonizing the now 31 national standards (this includes some European countries like Switzerland which are not member of the EU). The goal of harmonizing national standards into one unique standardization was guided by the wish to create common conditions for technical competition, to open a single trade market and remove trade barriers. The European market today brings more than 500 million people under its standardization framework.

To reach its goal, European standardization uses the so-called New Approach. Based on a decision of the European Parliament, the concept of technical harmonization in standardization was fixed in 1985 as follows:

- The European Directives set basic rules and framework conditions for technical requirements in specific fields (Machinery Directive, Low Voltage Directive, Electromagnetic Compatibility Directive and others).
- European Standardization sets fixed values for these basic requirements according to state-of-the-art technology.
- The application of European Standards conforms with European Directives and its application is voluntary.

The technical work is done by industry with no payment and the standardization organizations in CEN, CENELEC and ETSI organize the work and set the rules for writing standards open to anyone interested.

Once a European Standard (EN) is published the national members are required to transfer the EN into a national standard within a limited time. National standards that contradict it must also be withdrawn within a limited time. All members of EU and the European Free Trade Association (EFTA) must ensure that no national standardization work can be started, continued or published in the same technical field.

To organize standardization in Europe three organizations were founded: the European Standardization Committee (CEN), the European Standardization Committee for Electricity (CENELEC) and the European Standardization Committee for Telecommunication (ETSI).

The European Standardization Committee (CEN) is a private organization under Belgian law with its headquarters in Brussels, founded in 1991. It has 31 national members (28 EU and three EFTA), two consultant members (one European Commission and one from EFTA), seven associate members and four connected members (from Eastern Europe and Turkey). In CEN more than 60 000 experts work on more than 12 000 European standards. See www.cen. eu (accessed 13 February 2016) and also section 4.6.

The European Standardization Committee for Electricity (CENELEC) has the status of an international association under Belgian law with its headquarters in Brussels. It was founded on 13 December 1972 as the successor to the Comité Européen de Normalisation Eléctrique (CENEL) and later the Comité Européen de Normalisation Eléctrique Communication (CENELCOM). It has 33 national members in EU and EFTA and several thousand experts work on 6519 standards. See www.cenelec.eu (accessed 13 February 2016) and also section 4.5.

The European Telecommunications Standards Institute (ETSI) has direct members and is financed by member fees. Its structure is different from CEN and CENELEC. The only language in ETSI is English and the published standards are freely available to anybody. The public is connected to ETSI's work through national standardization organizations. ETSI has also members outside of Europe. See www.etsi.org (accessed 13 February 2016) and also section 4.7.

3

Standardization Processes

3.1 General

This chapter gives an overview of the way in which standardization processes are organized and structured. There are some basic rules that are applied in most leading organizations like the IEC, ISO, IEEE and regional bodies. On the other hand, each organization has its own detailed rules and principles, so, in practice, standardization work varies from organization to organization. Some large organizations like IEC and IEEE try to harmonize and coordinate their activities but in general there is only a little coordination between the different organizations in terms of what they do and how they do it.

Here information is given on the principles of the standardization process and work and in addition some details for the leading organizations.

3.2 Principles

Ten Rules

There are ten basic rules for the standardization process. The ten rules, which have been formulated by the Deutsche Kommission Elektrotechnik Elektronik Informationstechnik in DIN und VDE (DKE) (DIN 820), are listed in Box 3.1.

Voluntary participation in the standardization and applying standards to products and services is a basic requirement for the standardization process. The reason for this is to keep technical development open for innovation in case the standardization process is too slow.

'Public' means that everybody can get the information in draft documents. Usually there is a process for public participation and to enable the public to obtain the necessary information. 'Open to all interested parties' means that if someone is interested then there is a possibility

Practical Guide to International Standardization for Electrical Engineers: Impact on Smart Grid and e-Mobility Markets, First Edition. Hermann J. Koch.
© 2016 John Wiley & Sons, Ltd. Published 2016 by John Wiley & Sons, Ltd.

Box 3.1   Ten principles of standardization.

Voluntary
Public
Open to all interested parties
Consensus
Uniform and consistent
Relevant to the subject
Orientation on the state of the art
Orientation on the economic conditions
Orientation on the general benefit for society
To be international

---

for that person to participate in the standardization process. In some cases the interested person can participate directly or be represented by a professional organization.

'Consensus' means that the majority agrees with the content of standard text. Depending on the standard's development stage consensus is reached at 50%, 66% or 75% approval.

Consensus on the standard's text is important for the standard's wide use later on in the market. It is the goal for each standardization to reach 100% consensus. This is reached in many cases through technical discussions by the experts.

'Uniform and consistent' means that standards of one organization are all written in the same style. The organization needs to check the consistency and in case of conflict it should withdraw conflicting standards. 'Relevant to the subject' means that the discussion on the content is strictly focused on the relevant subject to avoid any discussions that are out of the scope of the work. This is necessary to finalize the work on the standard in a given time frame. 'Orientation on the state of the art' means that the standard's text is up to date with the relevant innovations and represents the highest quality possible at an economical cost. This is to keep standards current and it requires continuous revision of standards in a 3–8-year timeframe.

'Orientation on the economic conditions' means that requirements in standards must provide high-quality products for the user but also keep the cost within controlled limits to avoid overdesign.

'Orientation on the general benefit for society' means that the work of standardization experts should not be just for their benefit. The question 'does this standard bring any benefit to society?' needs to be answered positively.

'To be international' means that standards written for single nations or regions are not optimal as they cannot be applied at an international level and give benefit to all.

Rules for Expert Groups

There are some basic rules that are applied in all standardization organizations to allow good work on the standardization documents. These basic rules are shown in Box 3.2.

'The work is done by technical experts' means that the members of a working group are accepted experts in the technical field and they have experience of the related technology. Leading standardization organizations use an evaluation process to select the right experts for the work on the standards. This is important to deliver high-quality documents that will be accepted by the user.

```
┌─────────────────────────────────────────────────────────────────────────┐
│                  Box. 3.2    Basic rules for expert groups.               │
│  ─────────────────────────────────────────────────────────────────────   │
│  The work is done by technical experts                                    │
│  Experts are nominated by the industry or industrial associations.        │
│  Experts from all related technical fields shall be represented in the group. │
│  Results of the experts' groups shall be made public.                     │
│  Comments by interested members of the public must be considered by the expert group. │
│  In case of conflict between different opinions, an appeal to arbitration shall be made. │
│  The standardization organization needs to have rules and procedures in place. │
│  ─────────────────────────────────────────────────────────────────────   │
└─────────────────────────────────────────────────────────────────────────┘
```

'Experts are nominated by the industry or industrial associations' means that the experts are from industry and are paid by industry to bring their knowledge into the standardization work. This contrasts with the older days of government-paid standardization experts who are still active in some technical areas that do not require specific industrial knowledge and experience – for example, symbols in drawings.

'Expert groups from all related technical fields shall be represented in the group' means that it should be possible for all industries that would have to follow the proposed standard to participate in writing the standard. There must be a balance of experts in related technical fields – for example, manufacturers, users, test laboratories, consultants and authorities.

'Results of the expert groups shall be made public' means that there must be a process in place to inform members of the public about the standardization activities and the status of the document so that they can react and make their comments.

'Comments by interested members of the public must be considered by the expert group' means that there must be a process in place to obtain and consider comments. This needs to be documented and communicated from the expert group to the interested public for each comment.

'In case of conflict of different opinions an appeal to arbitration shall be made' means that there is a process in place for members of the public who make comments to ask for an independent evaluation group outside the expert group to resolve differences of opinion. This is to clarify the text before the standard is published.

'The standardization organization needs to have rules and procedures in place' means that it must be transparent and clear how the standardization process will work to solve conflicts.

Principal Working Rules

The everyday work in standardization must follow the principles shown in Box 3.3.

'Open process' means the standardization process is open to everybody. Each person must be able to propose a standardization and start a process after approval by an expert group of the standardization organization. If the proposal is declined, an explanation must be given.

'Participation of all interested persons' means that the expert group represents all interested persons or organizations. There are rules to handle the experts' participation and, in larger group, their representation by their technical organization. This participation is important for the benefit of the general public and society. A standardization plan for the whole process therefore needs to be established and all interested individuals are invited to participate. It is not acceptable if individuals gain an economic advantage.

'Public objection' means that before publishing the standard public comments must be evaluated. Individuals or entities delivering comments need to be invited to participate in the

```
┌──────────────────────────────────────────────────────┐
│                                                        │
│          **Box 3.3**   Principal working rules.        │
│        ────────────────────────────────────────        │
│                                                        │
│   Open process                                         │
│   Participation of all interested persons              │
│   Public objection                                     │
│   Public statement                                     │
│   Objection period                                     │
│   Arbitration procedure                                │
│   Method of operation                                  │
│   Consensus                                            │
│        ────────────────────────────────────────        │
│                                                        │
└──────────────────────────────────────────────────────┘
```

comment evaluation process to allow them to give explanations. If the individual or entity making the comments does not agree with the given evaluation, an arbitration process must be in place.

'Public statement' means that the standard document must be made available for comments to all interested persons before publication.

'Objection period' means that comments received must be processed in a given period of time. Typical timeframes are 3–6 months. Comment evaluation meetings must be organized in such a way that those giving comments are invited to the meeting or receive the documented decisions.

'Arbitration procedure' means that each individual or entity is allowed to ask for an appeal of his case by an independent group of experts.

'Method of operation' means that the standardization organization has procedures in place that allow the content of a standard to be created in a way that leads to mutual understanding and a common, accepted view. To reach this common understanding no formal voting is needed.

'Consensus' means that the content of the standard is accepted by all of the experts. There must be no remaining opposition to essential parts of the standard. The standardization process must try to bring all of the experts to a common understanding. This does not mean that unanimity is needed. In most standardization organizations two-thirds or 75% approval is required.

Composition of Standardization Groups

Standardization groups must have experts from all interested parties. The group size must reflect the needs of industry. In many cases groups of 15–20 experts cover this. In some cases, depending on the number of interested individuals, this number may be much larger. The largest had more than 1500 experts. In such cases, when a hundred or more experts are participating on the standardization work, procedures must be in place to manage the discussions within an acceptable timeframe.

Participating experts must have expertise and knowledge in the appropriate technical field and for international standardization they also need to know languages, in most cases English, and be able to deal with different nationalities and cultures. Experts must therefore be nominated by expert groups experienced in international standardization and must keep the experts and organizations that have authorized them informed. It is common for international and national standardization groups to invite guests to their meetings for information.

3.3   Legal Relevance

Technical standards are not laws. They give rules that are publicly available and free for everybody to use. Technical standards must be used in cases where this is required by law, by the authorities, or if they are part of a contract.

Technical standards are used in court cases involving failures and damages. Public authorities also use technical standards in court. In practice, standard users have a big advantage in the event of a failure as they can prove that the product has been manufactured according to the state of the art.

3.4   Benefits of Standardization

Companies often raise questions about the benefits of standardization versus the cost of experts participating in the standardization process. These questions are not easy to answer!

It is relatively easy to count the costs of an expert travelling internationally to standardization meetings over a period of some years but it is not so easy to count the benefits. On the other hand it is quite easy to calculate the costs if products do not conform to standards and redesign costs occur in order to meet standard requirements. Such costs are usually not small.

If a business's own experts participate, there are usually no redesign costs because of the business's own experts active participation in the standardization process and the opportunity to obtain information about coming technical changes. Here are some basic ideas on benefits of standardization:

* Safety. Electrotechnical products and solutions usually have safety concerns involving the risk of electrical shock. In many cases the electrical products also have risks associated with their potential to damage other equipment. Technical standards cover these safety aspects and give rules to avoid damage.
* International standardization opens markets and reduces trade barriers. Innovative products and solutions are more likely to be accepted if standards are in place. This enables users to trust such products and solutions when, as in most cases, they cannot prove the correctness of the statements of the manufacturer or solution provider. Contracting is simplified by using standards to fix the requirements to be met by products or solutions. This opens markets and new, innovative products and solutions enter the market much quicker.
* System compatibility. Technical standards enable products and solutions to be made compatible with other brands. This allows users to have a wider field of possible vendors and to establish competition for products and solutions. It reduces dependence on only one manufacturer.
* Tested products. Technical standards give requirements and design rules to be followed by products and solutions. Standards also offer rules on how to test quality with the standard. This guarantees a high quality level of the product or solution. In most cases the user would need external expertise for such evaluations and would have to bear high test costs.
* Fast distribution of innovative products and solutions. Technical standards are 'state of the art'. Standardization usually starts during the development phase of new products and solutions. The availability of technical standards helps products and services to gain acceptance from users and gives access to the market. This speeds up distribution, particularly when international standards are available.

- Communication platform. Technical standards and their standardization organizations play an important role in information exchange between companies, users, scientists, the general public and governments. A common view is generated based on transparent rules of cooperation established by the standardization organizations. This helps to create accepted technical positions on an international or national basis.

In 2000 the German standardization organization, DKE, estimated the benefit of standardization for Germany at a value of 16 billion euros per year (DIN 820).

4

Development of Standards

4.1 General

4.1.1 Basic Process

Processes for developing standards have some common steps and a lot of organization-specific details. The common steps are shown in Figure 4.1.

Proposal for a New Standard

Before a new standard can be written, there is a process created by the standardization organizations to evaluate the need for and benefit of this new standard versus the effort and cost to produce it and the impact on the market. In the IEC it is necessary for a minimum number of national committees (NCs) to approve the proposal and to send an expert to do the work. In IEEE there is a new standards standing committee under the control of the IEEE Standards Association Board to approve new work.

Draft Standard

In most standardization organizations it is required to present a draft standard document that shows the outline and main content of the document to be planned. The draft standards are used within the working group during the process of writing using various version numbers. At this stage the only control of the standard document is within the working group.

Draft Standard for Comments

After some progress is made in the working group and consensus is reached on major parts of the text, the draft standard is sent out for comments to interested individuals in the related technical field. In general working groups do not take too long before they send out a draft for

Practical Guide to International Standardization for Electrical Engineers: Impact on Smart Grid and e-Mobility Markets, First Edition. Hermann J. Koch.
© 2016 John Wiley & Sons, Ltd. Published 2016 by John Wiley & Sons, Ltd.

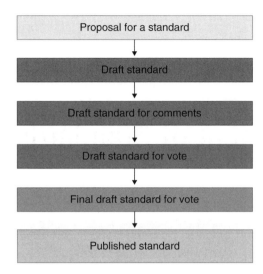

Figure 4.1   Common steps of standard development.

comments. Typical timeframes in major standardization organizations are 18 months up to 3 years, mostly depending on the subject and the market's need for the standard to be published. Experience shows that early circulation can bring feedback from the market on what is acceptable and what needs to be changed. In most standardization organizations this process of comments is formalized. In some cases more than a thousand comments have been received on one draft standard. According to the general rules each comment must be processed and evaluated and a decision made about possible changes to the text of the standard.

Draft Standard for Vote

This stage will be started when the working group is convinced that text is complete and could be published. Comments are expected but the general rule is that after the draft standard for comment has been circulated only editorial and light technical changes will be made. No major technical change of the text is expected, to avoid negative votes from those experts who contributed to the comment stage before.

Voting at this level in some organizations needs 50%, 66% or 75% approval. This is related to the type of document involved – for example, whether it is a standard, guide, or specification. With this voting result the decision is also made to go for the next stage of a final draft standard. The typical timeframe for this first voting on the standard is between 2 and 4 years from the starting date. In some standardization organisations only one step for voting is required, e.g. CENELEC.

Final Draft Standard for Vote

The final draft standard for vote is the last stage before the document is published. Possible technical changes at this stage are very limited. Only editorial changes and corrections to errors are accepted in general. The reason for this strictness in keeping the technical meaning of the text unchanged is that the previous steps gave the opportunity and time to change the technical content. The many experts who accepted the text might not be able to accept the text if technical changes are made.

The voting needs to reach at least two-thirds (e.g. 66% in IEC and 71% in CENELEC) approval in all major standardization organizations.

Published Standard

At this stage editorial work is done and writing-style checks are made to generate a document that looks the same as all the other standards of the related standardization organization. For leading standardization organizations like IEC, ISO or IEEE, professional editorial staff support this work. For the user and reader this formal appearance and writing style is very important for the correct understanding of the standard.

4.1.2   Organizations

International standardization organizations are the International Electrotechnical Commission (IEC) and the International Organization for Standardization (ISO).

Both are private entities located in Geneva, Switzerland. Telecommunication standards are published by the International Telecommunication Union (ITU), a suborganization of the United Nations.

The members of the IEC and ISO are NCs with one vote per country, independent of its size or economic strength. The major NCs are shown in Table 4.1.

The use of international standards from the IEC and ISO is not mandatory for national committees. National committees are free to decide to take or to leave IEC or ISO standards.

In the European Union the decision to adapt an international standard of the IEC and the ISO into a European standard (EN) is made at the European level at CEN for ISO and CENELEC for IEC. When the decision is made to transfer the standard to EN, with or without modifications, then each of the 28 member countries must transfer the EN into national country standards within a given time frame. CEN and CENELEC standards are seen as regional standards.

In North America the adaptation of IEC or ISO standards is not as strict as in Europe. It is more a decision by the user about which standards he will require. Here international standards are used in parallel with the North American standards of ANSI in the United States and the SCC in Canada. ANSI and SCC standards are seen as regional standards.

Table 4.1   Major national committees of the IEC and ISO.

Country	IEC NC	ISO NC
Germany	DKE	DIN
United States	ANSI	ANSI
France	AFNOR	AFNOR
United Kingdom	BSI	BSI
Spain	AENOR	AENOR
Italy	CEI	UNI
Russia	GOST	GOST
China	SAC	SAC
India	BIS	BIS
Japan	JISC	JISC
Canada	SCC	SCC

In Central and South America the IEC and ISO international standards are used depending on industrial influences. Where North American influence is greater, ANSI and SCC standards are used; where links to Europe are stronger, more IEC and ISO standards are applied.

A similar situation can also be found in Asia. China and India are mainly using standards from national sources, international sources such as the IEC and ISO and North American standards such as those of the IEEE, depending on the local industry and the authorities.

In India there is still a strong influence from British standards. This is seen with BSI standards, depending on the region and type of consulting company.

In Australia, the IEC and ISO are dominant, followed by IEEE standards in the electrical field. British standards from BSI may also be found, depending on the consultants active in the projects.

In Africa, the IEC and ISO dominate the standard requirements in general. In specific business fields, like the oil and gas business, North American standards are dominant.

On top of the international standardization organizations, the World Trade Organization (WTO) stands for the elimination of technical barriers to trade (TBT). The WTO requires national regulators to take international standards as a basis for their legal requirements and trade laws.

4.2   International Electrotechnical Commission (IEC)

4.2.1   General

The link to the IEC's web site is www.iec.ch (accessed 14 February 2016). The IEC is the oldest international standardization organization. It was founded in 1906 with support from nine countries. Today the IEC has 60 full members, which are nation states with their own national electrotechnical committees (NCs). There are 22 associate and 83 affiliate members, which have no voting rights and limited commenting rights and pay a reduced or no membership fee.

The IEC is a globally recognized organization to provide technical standards and related services, like conformity assessment of standards application. The goal of the IEC is to facilitate international trade in the electrotechnical field.

The founding countries were Canada, France, Germany, Italy, Japan, Sweden, Switzerland, the United States, and the United Kingdom – at this time the leading industrial countries of the world. Famous individuals belonged to the circle of founders, like Lord Kelvin, George Westinghouse and Werner von Siemens.

The IEC today is the world's leading standardization organization, which has prepared and published international standards for all electrical, electronic and related technologies since 1906.

4.2.2   Mission and Objectives

The mission of the IEC is

- To prepare and publish globally recognized international standards for all electrical, electronic and related technologies. These international standards serve as a basis for the development of national standards and as a reference when drafting international tenders and contracts.

• Through its members, the IEC promotes international cooperation on all questions of electrotechnical standardization and related matters such as the assessment of conformity to standards, in the fields of electricity, electronics and related technologies.

• The IEC charter embraces all electrotechnologies, including electronics, magnetics and electromagnetics, electroacoustics, multimedia, telecommunication, and energy production and distribution, as well as associated general disciplines such as terminology and symbols, electromagnetic compatibility, measurement and performance, dependability, design and development, safety and the environment.

The IEC's objectives are:

• to meet the requirements of the global market effectively;
• to ensure maximum worldwide use of its standards and its conformity assessment schemes;
• to assess and improve the quality of products and services covered by its standards;
• to establish the conditions for the interoperability of complex systems;
• to increase the efficiency of industrial processes;
• to contribute to the improvement of human health and safety;
• to contribute to the protection of the environment.

The IEC's founding documents proclaimed that: 'The future of science and particularly of electrical science is boundless.' This was its view in 1906 and we still cannot see an end to technical development.

4.2.3   Organization

The IEC is organized, in principle, on three levels. The membership level comprises all full-member national committees on the IEC Council and a number of national committees on the council board, including the IEC officers as IEC employees in the executive committee. This membership organization level is supported in its function and work by a central office with IEC employees and by the management advisory committees with elected industry volunteers.

The second organization level organizes and manages the IEC's work. The standardization management board (SMB) manages the work of producing standards by consensus based on its rules and processes. The market strategy board (MSB) watches the technology market and prioritizes the next activities for standardization work. The conformity assessment board (CAB) manages conformity assessment operations and systems.

The third level of organization is the working-group level, where the work is done. For the SMB this work is organized in technical committees (TC), technical advisory committees (AC) and sector boards (SB). The technical committees are technology oriented and organize their work in subcommittees (SC), working groups (WG) and maintenance teams (MT). The ACs are organized in SBs, advisory groups (AGs) and system committees (SysComs).

The WGs work on new standards, the MTs revise existing standards, the SBs prepare standardization in technical fields and the system committees deal with technical systems that have standards of equipment, services and devices involved belonging to several TCs or SCs.

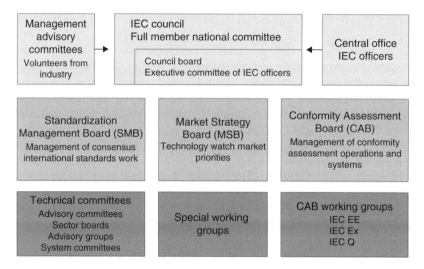

Figure 4.2   IEC organization in three levels.

The IEC, together with the ISO and ITU and coordinated with WTO rules, delivers international standards as a tool for barrier-free trade.

The MSB has special WGs and the Conformity Assesment Board (CAB) has three committees on Electrical Equipment (IEC EE), on explosive protection (IEC EX) and on quality (IEC Q). See Figure 4.2.

There are some specialities in the IEC. One principle is that IEC organizational decisions are only made by representatives of industry. The staff of the IEC only act to organize the standardization process and to support the experts in writing the standards.

Any leading position in the IEC is filled by experts from industry. This relates to the IEC president and any chairman or secretary of the technical committees and advisory boards. In this way the IEC wants to make sure that the standardization work follows topics that are relevant to the market.

It is the goal of the IEC's leadership to follow the technical work closely. Chairmen and secretaries play an important role in guiding the technical content of the standardization work.

The origin of the IEC in Europe still is reflected in the leading role of European national committees. Each nation has one national committee. So Switzerland or Denmark, for example, forms a national committee with one vote and one seat as does the United States or China.

More than 70% of the TCs or subcommittees SCs are European and more than 20% are from Germany, the country with the leading position in the IEC today, followed by the United States and France.

Work Structure

Standardization work is organized by TCs, SCs, WGs and MTs as shown in Figure 4.3.

Management Level

Management is formed by the council, where all the NCs that are members of the IEC have one seat and one vote. The standards management board (SMB) is the operational management with 15 members elected by the council. Seven members are permanent members due to their

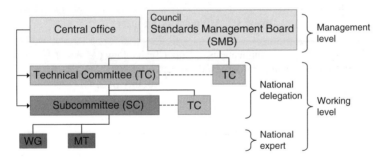

Figure 4.3 Work structure in the IEC.

leading position. Eight more members are elected. These eight positions change frequently. The 15 members and the chairman who is the Vice-President of IEC are elected by the Council.

Work Level

The work of standardization in IEC is split between national delegations and national experts.

Technical Committee

The TC and SC members are delegates from the NCs with full voting and commenting rights when permanent members (P members) or only commenting rights when observer members (O members). The TCs and SCs are the decision-making committees on any step in the standardization process.

National committees can send one or more delegates to the TC and SC meetings. If more than one delegate is sent, the NC needs to identify one chief delegate for the voting. Technical committees and SCs are led by a chairman and a secretary. The chairman is elected by all P members for a period of 3 years with the possibility of extension for a second 3-year period. The proposal for the chairman is made by the NC, which holds the secretariat. Each TC and SC is delegated to one NC, which nominates a secretary. The secretary is responsible for ensuring that the process of standardization of the TC or SC is correct. There is no time limit to the period for which secretaries may remain and usually secretaries stay for longer periods of time. This is important because they give stability to the standardization process by providing guidance to the WGs and MTs at the working level.

Working Group Level

The working level where the standards are written is the WG for new standards and the MT for existing standards. The WGs are started on basis of a new work item proposal (NWIP) after a positive vote from the NCs (more than 50% agree) and the nomination of a minimum of four experts for small TCs and SCs and five experts for large TCs and SCs to participate in the standardization work. The experts are nominated by NCs but act in the WG only on the basis of their technical expertise. They are not representing the opinion of the NC; they are only in close contact.

Once a standard is published and a revision is needed, the TC or SC will hold a vote. If there is a positive result (more than 50% of NCs agree) an MT will be established and experts from NCs will be nominated. In this case no minimum number of experts is required.

Box 4.1   Advisory committees.

Advisory Committee on Applications of Robot Technology (ACART)
Advisory Committee on Environmental Aspects (ACEA)
Advisory Committee on Electromagnetic Compatibility (ACEC)
Advisory Committee on Energy Efficiency (ACEE)
Advisory Committee on Safety (ACOS)
Advisory Committee on Security (ACSEC)
Advisory Committee on Electricity Transmission and Distribution (ACTAD)

The nominated experts of MTs do have the same status as those of WGs. They are members because of their expertise and do not represent the opinion of the NCs.

Advisory Committees (AC)

Beside the TCs in the IEC, several ACs are supporting the standardization work, see Box 4.1.

Advisory committees in the IEC support the standards management board (SMB). They prepare information for the SMB as a basis for decisions or they take over investigations formulated by SMB.

The members of ACs are representatives from related TCs and experts nominated by NCs. The general rules keep the membership small for better work efficiency. The recommended membership is 12 experts.

Advisory committees do have a strong impact on the standardization work of the TCs by working out recommendations to the SMB, which is then incorporated into a SMB decision and recommendation to the related TCs and SCs.

Advisory committees work outside the standardization processes of the TCs. They do not publish standards. Their work is limited to investigations to obtain information for recommendations to guide the work of TCs and SCs. Advisory committees are important for investigating specific topics, such as safety, future technical developments, or environmental protection, and to provide recommendations to the SMB. They invite experts from industry representing their related TCs and SCs or NCs. The typical result of their work is a report to be presented at the SMB meeting every 6 months. Advisory committees meet once or twice a year.

System Evaluation Groups (SEG)

System evaluation groups are a relatively new IEC structure introduced for the first time in 2014 to control more complex systems with more than one TC involved. The SEG is open to anyone and not just those who are NC experts. The task of the SEG is temporary – lasting 1 to 3 years – and involves preparing the launch of a system committee (SyC), which then has the right to release standards or to initiate standardization work in related TCs or SCs. Membership in a SyC is only possible through NCs.

Today there are several SEGs active and some have already been transferred into SyC – see Table 4.2.

SEG 1 Smart Cities

System Evaluation Group 1, Smart Cities, was a German and Japanese initiative. It covers the electrical aspects of power generation, consumption and all the various possibilities of digital control and management, including those involving the Internet. It will produce basic

Table 4.2 System evaluation groups.

Name	Title	Active
SEG 1	Smart Cities	since 2014
SEG 2	Smart Grid	transferred to SyC Smart Energy 2015
SEG 3	Ambient Assisted Living	transferred to SyC Active Assisted Living in 2015
SEG 4	Low Voltage Direct Current Applications, Distribution and Safety for use in Developed and Developing Economies	since 2014
SEG 5	Electrotechnology for Mobility	since 2015
SEG 6	Non-traditional Distribution Networks/Microgrids	since 2015

Table 4.3 Working groups of SEG 1 Smart Cities.

WG 1	City Service Continuity
WG 2	Urban Planning and Simulation System
WG 3	City Facilities Management (CFM)
WG 4	Use Case – Smart Home
WG 5	Use Case – Smart Education
WG 6	Smart Cities Assessment
WG 7	Standards Development for Smart Cities using the City of Johannesburg
WG 8	Mobility and Logistics
WG TG 1	Inventory of relevant existing standards
WG TG 2	Reference architecture – Generic use cases
WG TG 3	Roadmap

information about possible architecture models of Smart Cities and a standardization road map for any TC and SC involved.

Today, SEG 1 has more than 200 members worldwide, split into more than ten working groups as shown in Table 4.3.

System Evaluation Group 1 Smart Cities held its first meeting in January 2014 in Berlin, Germany, where the WGs shown in Table 4.3 were founded. The first technical committee meeting was held in March 2015 in London. The work is in progress and the first draft documents for comments are expected for 2016 and 2017.

SEG 2 Smart Grid

System Evaluation Group 2 started in 2012, working on the impact of changing electric generation, power transmission and distribution based on renewable energies and the new possibilities of digital control and management including opportunities presented by the Internet. IEC 61850 is a basic standard for Internet protocols and data exchange. This standardized data exchange allows manufacturers to develop technologies that are able to communicate on the basis of the IEC 61850 standard series. Many new ideas for operating a future power network are now available and they function. Communication is independent from manufacturers' individual solutions. It is standardized for interchangability.

In preparation for SEG 2, the scope and roadmap for the Smart Energy SyC were created in 2015 – see below.

SEG 3 Ambient Assisted Living

System Evaluation Group 3 started in 2013 to provide some initial explanations regarding how technology can help older people to manage their daily living in their private homes. In a roadmap, SEG 3, showed the standardization work that needed to be done. In 2015, SEG 3 was transferred into Active Assisted Living (SyC) – see below.

SEG 4 Low Voltage Direct Current

System Evaluation Group 4 Low Voltage Direct Current Applications, Distribution and Safety for use in Developed and Developing Economics was founded in 2015 on the new principle of using DC electric energy and transforming it to AC for transmission and distribution. Most renewable sources, such as photovoltaic, are generated as DC and may be used directly, for example for light, cooling, electromobility. System Evaluation Group 4 will work out requirements in standardization to set up a roadmap. More than 100 experts have been nominated to SEG 4, who are working in six WGs as shown in Table 4.4

The first meeting of SEG 4 was held in Princeton, United States, in May 2015. The WGs were established as shown in Table 4.4.

SEG 5 Electrotechnology for Mobility

System Evaluation Group 5 Electrotechnology for Mobility investigates the interactions between plug-in electric vehicles and the supply infrastructure. It covers the aspects of safety, interoperability and system performance for future standards. There are interactions with SEG 2 Smart Cities, manufacturers, suppliers, ISO TC 22 and several related TCs. A roadmap will include the standardization activities and the interactions. SEG 5 started in 2015.

SEG 6 Nontraditional Distribution Networks/Microgrids

Nontraditional distribution networks and micro grids are independent electric power-generation and consumption islands connected to public power supply system only for times of power shortage on the power island. The so-called micro grids use different sources of power generation like photovoltaic, wind, water, biomass and storage systems like pump storage or batteries. They use computer and digital control to manage the island network stability by interactions with demand control and storage control and maximum use of the various generation sources. The first experiences worldwide have been very positive and fast market growth can be seen.

Table 4.4   Low voltage direct current WGs.

WG 1	Current Status; Standards and Standardization
WG 2	Stakeholder Assessment and Engagement
WG 3	Market Assessment
WG 4	Collection and Rationalization of Current Voltage Data – LVDC
WG 5	Collection and Rationalization of LVDC Safety Data
WG 6	LVDC for Electricity Access

Table 4.5   System committees.

Name	Title	Active
SyC AAL	Active Assisted Living	since 2015
SyC Smart Energy	Smart Energy	since 2015

Table 4.6   Working Groups of SyC AAL.

WG 1	User Focus
WG 2	Architecture and Interoperability
WG 3	Quality and Conformity Assessment
WG 4	Regulatory Affairs

System Evaluation Group 6 was initiated by China and at the time of writing had about 30 expert members. Three working groups had been established:

- WG 1: situation assessment;
- WG2: use cases;
- WG3: specific needs for micro grid technology standardization.

A first meeting was held in 2015 in Clamart, France.

System Committees

The SyCs (see Table 4.5) are to coordinate standardization activities in related areas. The SyC only has members that are nominated by NCs and has the right to write and publish standards. The coordination work of the SyC includes any Technical Committee releasing standards in the related area. The SyC may propose new standardization activities in the related TCs or SCs or may release a standard document by himself.

SyC AAL Active Assisted Living

This SyC AAL is working on developing a vision of active assisted living for the market including proposals for interoperability and product accessibility. It communicates requirements for standardization work to the related product committees. At the time of writing, SyC AAL had 24 NCs as members and had started work in four WGs as shown in Table 4.6.

The work of WG 1 is focused on all user-related issues related to liability, safety, functionalities, security and privacy. In WG 2, interoperability is the main work of the products available with a common architecture. In WG 3 work focuses on the criteria for quality, testing, tools and standards. WG 4 covers aspects of regulation of the products used for assisting people in their private surrounding.

At the time of writing work had started on definitions for the International Electrotechnical Vocabulary of expressions related to active assisted living. The first meeting was held in March 2015 in Frankfurt, Germany.

Table 4.7   Working groups of SyC smart energy.

WG 5	Methodology and Tools
WG 6	Generic Smart Grid Requirement

SyC Smart Energy

The smart energy SyC covers the standardization aspects of the smart grid and smart energy requirements related to renewable electric energy and the use of digitalization including the Internet.

It also covers aspects of interaction with heat and gas. There are strong links to SEG 1 Smart Cities. At the time of writing there were about 30 NC members and the work started in June 2015 in Beijing, China, with two working groups – see Table 4.7.

In WG 5 methodology and tools are provided for standardization with process definitions, templates, actor lists, XML schemes and use cases. In WG 6 a series of projects will be set up to learn about the requirements of smart energy to prepare the standardization.

Technical Committees (TCs and SCs)

The technical work in IEC is split between TCs and SCs. Technical committees and SCs are led by a chairman and a secretary. The chairman is responsible for guidance on the technical content of the TC or SC and will lead the TC and SC meetings. The secretary is responsible for the ensuring that the correct process is followed in the standardization work carried out by WGs and maintenance teams (MTs).

The TCs are numbered chronologically, so TC1 is the oldest one. At the time of writing the latest one was TC 122, established in 2014. There is a total of 174 TCs and SCs, which operate more than 700 WGs and MTs. Table 4.8 gives an overview.

4.2.4   Members

In the IEC, the only members are NCs. They have the right and duty to lead a TC by taking the responsibility of the Secretariat. The Secretariat is linked to the TC as long as the TC exists and is led by a person nominated by the NC and has to follow the rules of the IEC Directives.

The members of IEC today comprise about 98% of all the nations of the world. They have different statuses: P members have full voting rights; O members have no voting rights and may only observe and participate in meetings; and affiliate members who are informed about IEC activities. The members of management are elected by the NCs to manage the IEC; the members of WGs, MTs and task forces (TF) only can be nominated by NCs, as stated in the IEC Directives.

This makes the IEC a closed society with nominated experts and delegates from the NCs. Only recently has it been possible to be a member of the System Evaluation Group (SEG) without nomination through the NC as an interested person.

A new public commenting and participation rule will allow participation in a standardization process through an Internet link without the nomination of the Committees. The possibility of public comments on standards during their development process made it necessary to connect this channel of comments to the NC as the first point of entry. The NCs are then responsible if the comments are passed on to the respective WG.

Table 4.8 Overview of TCs.

Committee	Title
TC 1	Terminology
TC 2	Rotating Machinery
TC 3	Information Structures and Elements, Identification and Marking Principles, Documentation and Graphical Symbols
TC 4	Hydraulic Turbines
TC 5	Steam Turbines
TC 7	Overhead Electrical Conductors
TC 8	Systems Aspects for Electrical Energy Supply
TC 9	Electrical Equipment and Systems for Railways
TC 10	Fluids for Electrotechnical Applications
TC 11	Overhead Lines
TC 13	Electrical Energy Measurement and Control
TC 14	Power Transformers
TC 15	Solid Electrical Insulating Materials
TC 17	High-Voltage Switchgear and Controlgear
TC 18	Electrical Installations of Ships and of Mobile and Fixed Offshore Units
TC 20	Electric Cables
TC 21	Secondary Cells and Batteries
TC 22	Power Electronic Systems and Equipment
TC 23	Electrical Accessories
TC 25	Quantities and Units
TC 26	Electric Welding
TC 27	Industrial Electroheating and Electromagnetic Processing
TC 28	Insulation Co-ordination
TC 29	Electroacoustics
TC 31	Equipment for Explosive Atmospheres
TC 32	Fuses
TC 33	Power Capacitors and their Applications
TC 34	Lamps and Related Equipment
TC 35	Primary Cells and Batteries
TC 36	Insulators
TC 37	Surge Arresters
TC 38	Instrument Transformers
TC 40	Capacitors and Resistors for Electronic Equipment
TC 42	High-Voltage and High-Current Test Techniques
TC 44	Safety of Machinery – Electrotechnical Aspects
TC 45	Nuclear Instrumentation
TC 46	Cables, Wires, Waveguides, RF Connectors, RF and Microwave Passive Components and Accessories
TC 47	Semiconductor Devices
TC 48	Electrical Connectors and Mechanical Structures for Electrical and Electronic Equipment
TC 49	Piezoelectric, Dielectric and Electrostatic Devices and Associated Materials for Frequency Control, Selection and Detection
TC 51	Magnetic Components and Ferrite Materials
TC 55	Winding Wires
TC 56	Dependability
TC 57	Power Systems Management and Associated Information Exchange
TC 59	Performance of Household and Similar Electrical Appliances
TC 61	Safety of Household and Similar Electrical Appliances
TC 62	Electrical Equipment in Medical Practice

(Continued)

Table 4.8 (*Continued*)

Committee	Title
TC 64	Electrical Installations and Protection against Electric Shock
TC 65	Industrial-Process Measurement, Control and Automation
TC 66	Safety of Measuring, Control and Laboratory Equipment
TC 68	Magnetic Alloys and Steels
TC 69	Electric Road Vehicles and Electric Industrial Trucks
TC 70	Degrees of Protection provided by Enclosures
TC 72	Automatic Electrical Controls
TC 73	Short-Circuit Currents
TC 76	Optical Radiation Safety and Laser Equipment
TC 77	Electromagnetic Compatibility
TC 78	Live Working
TC 79	Alarm and Electronic Security Systems
TC 80	Maritime Navigation and Radiocommunication Equipment and Systems
TC 81	Lightning Protection
TC 82	Solar Photovoltaic Energy Systems
TC 85	Measuring Equipment for Electrical and Electromagnetic Quantities
TC 86	Fibre Optics
TC 87	Ultrasonics
TC 88	Wind Turbines
TC 89	Fire Hazard Testing
TC 90	Superconductivity
TC 91	Electronics Assembly Technology
TC 94	All-or-Nothing Electrical Relays
TC 95	Measuring Relays and Protection Equipment
TC 96	Transformers, Reactors, Power Supply Units, and Combinations Thereof
TC 97	Electrical Installations for Lighting and Beaconing of Aerodromes
TC 99	System Engineering and Erection of Electrical Power Installations in Systems with Nominal Voltages above 1 kV a.c. and 1.5 kV d.c., particularly concerning Safety Aspects
TC 100	Audio, Video and Multimedia Systems and Equipment
TC 101	Electrostatics
TC 103	Transmitting Equipment for Radiocommunication
TC 104	Environmental Conditions, Classification and Methods of Test
TC 105	Fuel Cell Technologies
TC 106	Methods for the Assessment of Electric, Magnetic and Electromagnetic Fields associated with Human Exposure
TC 107	Process Management for Avionics
TC 108	Safety of Electronic Equipment within the Field of Audio/Video, Information Technology and Communication Technology
TC 109	Insulation Co-ordination for Low-Voltage Equipment
TC 110	Electronic Display Devices
TC 111	Environmental Standardization for Electrical and Electronic Products and Systems
TC 112	Evaluation and Qualification of Electrical Insulating Materials and Systems
TC 113	Nanotechnology Standardization for Electrical and Electronic Products and Systems
TC 114	Marine energy – Wave, Tidal and other Water Current Converters
TC 115	High Voltage Direct Current (HVDC) transmission for DC Voltages above 100 kV
TC 116	Safety of Motor-Operated Electric Tools
TC 117	Solar Thermal Electric Plants
TC 118	Smart Grid User Interface
TC 119	Printed Electronics
TC 120	Electrical Energy Storage (EES) Systems
TC 121	Switchgear and Controlgear and their Assemblies for Low Voltage
TC 122	UHV AC Transmission Systems

Table 4.9 Types of IEC documents.

Abbreviation	Name	Type
IS	IEC International Standard	Consensus and without any contradiction to other standard
TS	IEC Technical Specification	No contradiction to IS
TR	IEC Technical Report	No mandatory requirements
Guides	IEC Guides	Rules to write standards
PAS	IEC Publicly Available Specification	No contradiction to IS, limited time
ITA	IEC Industry Technical Agreement	outside the IEC process
TTA	IEC Technology Trend Assessment	Position paper on technology development

4.2.5   Types of Documents

In the IEC there are seven different types of documents to cover the different requirements of standardization. An overview is given in Table 4.9.

International Standard (IS)

An international standard (IS) is a document that is based on the consensus of the experts nominated by the NCs. It is a normative document with mandatory requirements that have been approved by the NCs based on a public inquiry. Approval requires a two-thirds majority of votes and less than 25% negative votes.

Technical Specification (TS)

A TS is a document that covers technical subjects that are still under development or which show insufficient consensus because of missing knowledge or experiences. A TS is normative in nature and is approved by two-thirds of the votes of NCs. The final vote takes place when the committee draft for vote (CDV) has been circulated to the NCs. The final goal for any TS is the IS.

Technical Report (TR)

A TR is information worked out by TCs or SCs, which is not meant to be an IS. It is only meant to be published for information with no mandatory requirement. The content is only informative. Approval requires only a simple majority of the NCs.

Guides

Guides in the IEC give information about how to write a standard. Today there are about 50 different guides available in the IEC. There are three guides that have a normative character:

- *IEC Guide 104 Safety*
- *IEC Guide 107 Electromagnetic compatibility*
- *IEC Guide 108 Application of horizontal standards*

The guide on safety gives information on the preparation of standards covering safety aspects. This includes the safety of people, domestic animals, livestock and property.

The guide on electromagnetic compatibility (EMC) gives information for product standards when chapters on EMC are planned. It gives procedures to ensure that these publications are consistent with current practice and to avoid overlapping with existing documents.

The guide on the application of horizontal standards defines and establishes the relationship between horizontal and product standards. Horizontal standards cover requirements (for example, safety, environment, ratings) that are used in product standards.

Publicly Available Specification (PAS)

The PAS is a document that has reached consensus within an IEC WG or a WG outside the IEC. Its requirements do not conflict with existing international standards and it has a limitation of 3 years before it will be withdrawn. The reason for a PAS is to inform industry about the planned standardization to provide confidence in the industrial development processes. A simple majority of the TC or SC is required for approval.

Industry Technical Agreement (ITA)

The document of an ITA is worked out by industry outside the IEC and will be distributed for information only.

Technology Trend Assessment (TTA)

The TTA document is worked out by TCs or SCs to provide information for the standardization process on how technology develops and where new standardization activities are needed.

In addition to the stand-alone standard documents, there are three types of additional documents as shown in Table 4.10.

Amendments

The amendments are meant to add mandatory text to an existing standard following the consensus process in the same way as for the standard. They are used to add text to a standard without opening a complete revision process to the standard. A standard should not have more than two amendments.

Technical Corrigenda

The technical corrigenda are used to correct failures in the standard text. Technical corrigenda are immediately valid and are published and listed directly. The standards user must check if any corrigendum has been released for a standard being used. There are more than 1000 technical corrigenda available in IEC.

Table 4.10   Additional standard documents.

Name	Type
Amendment	To add information to an International Standard
Technical corrigenda	To correct failures
Interpretation sheets	To explain standard requirements

Interpretation Sheets

The interpretation sheet is used to give some specific information on standard requirements to aid understanding. They must be short: one or two pages. Today there are about 50 interpretation sheets published and listed.

4.2.6   Dresden Agreement

4.2.6.1   General

European countries are strong participators in IEC standardization work and hold leading positions in the IEC. European standardization work under the rules of CEN and CENELEC is strongly internationally orientated. For this reason the boards of CEN and CENELEC check each new standardization task with the IEC or ISO. Only if the IEC or ISO denies writing a standard will a working group be established in CEN or CENELEC.

To improve the coordination of standardization activities between CENELEC and IEC further, the Dresden Agreement went into effect in September 1996. The goal of the improvement was to synchronize the commenting and voting cycles of IEC and CENELEC so that, in the end, the IEC and EN documents can be published at the same time and, if possible, without any difference in content and without an extended time schedule.

4.2.6.2   Processes

The process of the Dresden Agreement is shown in Figure 4.4.

The parallel voting process on IEC and EN standards starts with the proposal to write such a standard in the IEC. The right to bring in a proposal on a new standard is with the IEC National Committee or the IEC or CENELEC TC or SC. Once the proposal has passed the IEC approval criteria of more than 50% approvals by NCs and a minimum of four experts nominated to be active members of the new WG, the WG can start its work. In cases when the total number of full NC members with voting right (P members) is larger than 16, a minimum of five experts are needed to start the work in the WG. If the criteria are not met, no work will be started. The IEC is very strict on that because it proves the market relevance of a planned standard.

This requirement should reflect that there is a market need for the proposed standard expressed by the industry sending a minimum of four or five experts to the WG.

Once the WG has started work and a committee draft (CD) is available for comments, this document is circulated only in the IEC. Only after the content of the document has reached a more stable status and the WG with the IEC TC decides to circulate the document as a CDV will the document be circulated to the IEC and CENELEC NCs.

In the IEC, at this stage, the NCs make technical or editorial comments on the content and vote on the document. The voting must reach two-thirds approval and must not have more than 25% disapprovals before the next stage in IEC can be reached. In CENELEC, at the CDV stage of the document, only comments are collected from NCs but no formal vote on the document is taken; CENELEC only votes on the final draft of document, which is provided by the IEC after the CDV has been approved in a vote and the comments from the NCs have been worked into the document.

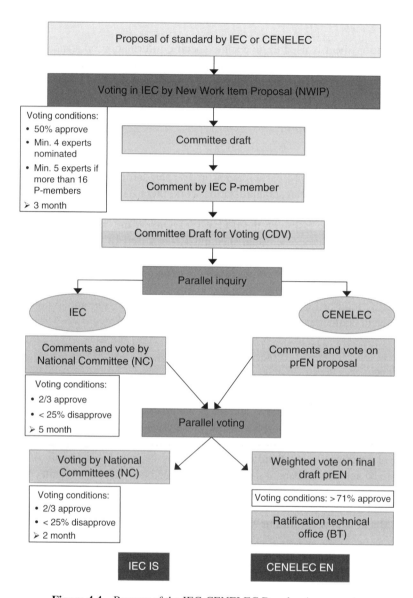

Figure 4.4   Process of the IEC-CENELEC Dresden Agreement.

The final draft in the IEC is circulated as 'final draft international standard' (FDIS) for a formal vote without the possibility of NCs giving technical comments. Here, again, a two-thirds approval and less than 25% disapproval is the criterion to meet. In CENELEC the weighted voting system applies and a majority of 71% is required to pass.

In the IEC the document will be published right away; in CENELEC the technical office (BT) will ratify the new standard and set a timeframe for the CENELEC member countries to adopt the standard and withdraw conflicting national standards.

Table 4.11   Steps in standards.

Proposal stage	Covers the proposal and the voting process on this new work NP = new work-item proposal
Preparatory stage	Covers the start of the new work until the first draft for circulation to NCs
Committee stage	Covers the first circulated draft to NCs for receiving comments CD = committee draft
Inquiry stage	Covers the voting draft document of the working group or maintenance team for comments and vote by NCs CDV = committee draft for vote
Approval stage	Covers the approval by NCs without further technical comments FDIS = final draft international standard
Publication stage	Covers the IEC editorial work on the standard for publication, which has been voted for and approved IS = international standard

4.2.7   Steps in Standards

The IEC standardization processes follow, in principle, the steps as shown in Table 4.11.

Proposal Stage

During the proposal stage, information about a new standardization project is collected and the possible title and the scope are prepared. The proposals for new work may come from NCs, the TC or one of its WGs or MTs. The TC may install an ad hoc group of experts to prepare the new standard work or it may ask an NC to work out the proposal.

The proposal of the new work must provide a title, a scope for the proposed work, a content description and a first-draft document to cover the main topics of the planned new standard.

The secretary of the TC preparing the proposal documents needs to identify the convenor by name and address, which is a mandatory requirement of the proposal form.

The IEC will not start new work without having a detailed definition of the content of the new standard and who will lead the WG as convenor.

The process to close the proposal stage is the new work item proposal (NP). The NP is sent to the member NCs of the TC for approval and the nomination of experts.

The acceptance requirements of the NP are as follows:

- Criterion 1: a simple majority of the full voting members (P members of the related TC).
- Criterion 2: a minimum of four nominated experts to the new WG for TCs with 16 or less full members; for 17 and above the requirement of experts is a minimum of five.

Usually the first criterion is easy to pass but the second is harder because NCs need to nominate experts who are willing to participate in the work and to attend the meetings – and the nominated experts must come from four or five different NCs.

Once the criteria have been met, the new WG will be established, receiving the next free number of the TC's sequence.

A tight timeframe of 3 years in total is given to the WG for finishing the work.

Preparatory Stage

The preparatory stage covers the working time to prepare the first draft of the document by the experts of the WG. Depending on quality of the entry document presented with the NP, the WG should be ready after 6 to 12 months with a first draft for circulating to the NCs.

The document is then called a committee draft (CD) and is meant to collect comments from the NCs to get a first feeling of whether the content will be acceptable and could find consensus. The document's language is English only at this stage.

Committee Stage

At the committee stage the NC's technical comments are worked into the document. The time for collecting comments can be set at 2, 3 or 4 months and the CD can be sent for comments several times. Two cycles of comments are typical, sometimes three, but seldom four. After receiving the comments the secretary will prepare a document together with the WG with a decision made for each comment, stating if it was accepted or not and why. A 4-week timeframe is given for this so-called 'observation of the Secretariat'.

In cases for clarification with the NCs the comments are the subject of discussion at the next TC meeting. The discussion must follow the requirements of consensus.

The committee stage ends when consensus is found on any technical topic of discussion.

Inquiry Stage

During the inquiry stage the document is circulated for a vote as a CDV. This is the last chance for technical comments. The votes are submitted by the NCs as positive, negative or abstention. Abstentions are not counted in the voting evaluation; these are lost votes. To pass the vote, the following criteria need to be fulfilled:

- Criterion 1: two-thirds majority of the NCs voting positive or negative.
- Criterion 2: no more than a quarter of the total number of votes should be negative.

Comments received are classified as accepted or not accepted with the observations of the secretariat within 4 weeks. This is the same as in the committee stage. After 3 months following the voting, the final text must be ready for the approval stage.

Approval Stage

The approval stage does not allow any more technical comment on the FDIS. Only corrections to errors or improvements to the text's style are allowed at the editorial level. The NC can only vote positive or negative. A negative vote is only chosen if the NC finds the standard unacceptable.

The approval stage again has two criteria to pass:

- Criterion 1: a two-third majority must cast a positive vote.
- Criterion 2: not more than one-quarter of the votes may be negative.

With a positive vote, the next and last stage in the standardization process is reached – the publication stage. At this stage a French version of the text will be provided within 2 months.

Table 4.12 Timeframe of documents in the standardization process.

NP	New work item proposal stage	3 months
	Proposal stage	
CD	Committee draft	2, 3 or 4 months
	Committee stage	
CDV	Committee draft for vote	3–5 months
	Inquiry stage	
FDIS	Final draft international standard	2 months
	Approval stage	
IS	International standard	
	Publication stage	

Publication Stage

At the publication stage the FDIS is transferred from the responsibility of the WG of the TC to the IEC central office editorial staff to prepare the published version. The IEC central office will check the document's style, update all the references and bring in line the English and the French versions of the text. Publication of the new standard will take place within 1 to 1.5 months.

The publication stage ends with the publication of the standard with the year and month indicated. No transition time is given to old standards; they are replaced immediately with the publication of the new standard. The timeframe of the different documents of the standardization process is given in Table 4.12.

The new work item proposal (NP) has a 3-month voting period. The result of the voting is documented by the result of voting on NP (RVN).

The committee draft (CD) has a 2-, 3- or 4-month voting period. The timing will be chosen by the WG together with the secretary, depending on the urgency and complexity of the document. The comments will be documented with the compilation of comment (CC) on the CD.

The CDV has a 3–5 month voting period. The voting results and the comments are published as the result of voting on the CDV (RVC).

The FDIS has a 2-month voting period. The voting result is published as the report of voting on an FDIS (RVD).

Maintenance

The IEC standards receive a fixed maintenance date once they are published. The maintenance may be between 3 and 10 years, depending on the content and the need to adjust the content to ongoing developments.

Once the expected maintenance date is reached, the review report is circulated to NCs by the TC secretary to propose a review of the document or an extension of the maintenance date.

4.3   International Organization for Standardization (ISO)

4.3.1   General

The ISO is a nongovernmental organization based in Geneva, Switzerland, where it was founded in 1947. It represents worldwide national standardization organizations similar to the IEC. Membership in the ISO is only open to national authorized standardization

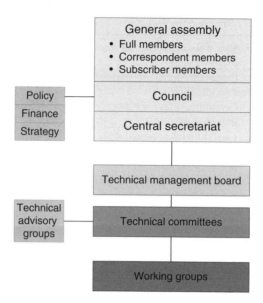

Figure 4.5   Structure of the ISO.

organizations, which represent the opinion of the whole country in the ISO. The ISO is therefore like a bridge between public sector national representation and private sector industry. The ISO network today connects 162 countries. More than 50 000 experts are working on a total of 16 500 international standards.

The goal of the ISO is to develop and support standardization work and support any activity related to that goal on a global basis. Its standardization is seen as a key element in encouraging and supporting free trade between countries and the exchange of any goods or services.

The ISO is cooperating with more than 600 international organizations worldwide. Its structure is shown in Figure 4.5.

The general assembly consists of full members or member bodies, corresponding members and subscriber members. Full members or member bodies have voting rights and are actively participating in ISO TCs and WGs. Corresponding members have observer status in the technical work and in strategic groups. They do not have voting rights but participate at meetings. Subscriber members are involved on an internal basis and cannot participate in meetings. At the time of writing, ISO listed 162 full members and 119 member bodies, 38 corresponding members and five subscriber members.

The council of the ISO governs the organization. It is led by a president and consists of 20 member bodies with full voting rights.

The council meets twice a year and the membership rotates to ensure that it is representative of the member countries. The council is the operative management of the central secretariat. Its role includes several supporting functions related to conformity assessment, consumer issues, developing country matters, policy, finances and strategy.

The technical management board (TMB) is responsible for leading the TCs through the standardization process. The technical work on the standards is carried out by the working group with experts.

There are more than 250 TCs covering a wide spectrum of interests, like TC 1 for Screw Threads, TC 11 Boiler and Vessels, TC 22 Road Vehicles, TC 74 Cement and Lime, TC 100 Chains, TC 150 Implants for Surgery, TC 192 Gas Turbines, TC 217 Cosmetics, TC 255 Bio Gas and TC 296 Bamboo and Rattan, just to mention a few.

The ISO has three types of membership. A member is a fully authorized member with voting, electing and commenting rights. Members pay a fee, which is related to the number of TCs in which they are active. Members are also required to participate actively in ISO. If they fail to carry out their duties of voting, commenting, electing or participating in TC meetings their membership rights will be withdrawn and they will be given a corresponding status with fewer rights and, of course, lower fees. This rule is similar to the position taken by the IEC and prevents inactive members who are on the list but do not contribute.

A corresponding member has the right to participate in meetings and comment on documents but is not allowed to vote. Subscriber members are connected to the ISO on an informal basis. They do not have rights to participate in standardization work and they do not have fees.

The council is elected by the members and is the management board of the ISO. The council is supported by the central secretariat for daily business operation. The central secretariat is the heart of ISO, coordinating the technical work with the voting and commenting processes. Technical officers of the TMB take care of handling standards, voting on standards and any comment coming from NCs. The TCs cover various fields of technology. They are led by a chairman and a secretary. The work of standards is carried out by WGs. Working groups are assembled through nomination of experts by the NCs.

The ISO publishes international standards in any field except in electrotechnology. The standardization process in ISO is very similar to the IEC process.

In several technical cases, mainly when electrotechnology meets mechanical engineering, the ISO and IEC work in so-called joint ISO/IEC committees.

4.3.2 ISO/IEC Joint Technical Committee (JTC 1)

The Joint Technical Committee JTC 1 covers aspects of information and communication technology (ICT). This TC covers the field of ICT of tools and systems to capture, process, secure, transfer, interchange, present, organize, store and manage information.

There are about 20 subcommittees covering different aspects of ICT, as shown in Table 4.13.

There are more than 500 standards published under JCT 1, covering the wide domain of information and communication technology used in the electrotechnology field of IEC and in any other area of ISO. The standards are published using an ISO/IEC number of the 40 000 series.

4.3.3 ISO Strategy

The strategic goals of ISO are as listed in Box 4.2.

The ISO is entitled to be the one international and global body to provide a collection of multisector standards to cover any field of technology except electrotechnology. The standards released by ISO are written to unique rules and provide a common understanding.

To involve any interested party in the standardization process is an important and to provide standards that are widely used are important goals. This openness is sometimes not easy to

Table 4.13   Subcommittees of JCT 1.

ISO/IEC JTC 1/SC 2	Coded Character Sets
ISO/IEC JTC 1/SC 6	Telecommunications and Information Exchange between Systems
ISO/IEC JTC 1/SC 7	Software and Systems Engineering
ISO/IEC JTC 1/SC 17	Cards and Personal Identification
ISO/IEC JTC 1/SC 22	Programming Languages, their Environments and System Software Interfaces
ISO/IEC JTC 1/SC 23	Digitally Recorded Media for Information Interchange and Storage
ISO/IEC JTC 1/SC 24	Computer Graphics, Image Processing and Environmental Data Representation
ISO/IEC JTC 1/SC 25	Interconnection of Information Technology Equipment
ISO/IEC JTC 1/SC 27	IT Security Techniques
ISO/IEC JTC 1/SC 28	Office Equipment
ISO/IEC JTC 1/SC 29	Coding of Audio, Picture, Multimedia and Hypermedia Information
ISO/IEC JTC 1/SC 31	Automatic Identification and Data Capture Techniques
ISO/IEC JTC 1/SC 32	Data Management and Interchange
ISO/IEC JTC 1/SC 34	Document Description and Processing Languages
ISO/IEC JTC 1/SC 35	User Interfaces
ISO/IEC JTC 1/SC 36	Information Technology for Learning, Education and Training
ISO/IEC JTC 1/SC 37	Biometrics
ISO/IEC JTC 1/SC 38	Cloud Computing and Distributed Platforms
ISO/IEC JTC 1/SC 39	Sustainability for and by Information Technology
ISO/IEC JTC 1/SC 40	IT Service Management and IT Governance

Box 4.2   Strategic goals of ISO.

1. Establish worldwide accepted standards.
2. Involve any interested party.
3. Open to any partner to work out standards.
4. Support standards in developing countries.
5. Support the use of ISO standards instead of regional regulations.
6. Neutral source of standards.
7. Provide an efficient standardization process.

---

handle because of a wide span of interests in the standardization work. With a transparent standardization process and with discussions focused on the technical content, experience shows that well accepted standards are provided.

In developing countries, often the organizational structure is not well developed and the availability of experienced experts is low. The ISO has developed support functions to give help to participants from developing countries and to make ISO standards available to their industries.

The principle of international standards is to establish technical rules on a global basis to support international trade and an exchange of goods and services. The ISO supports the use of standards instead of regional and national regulations, which are a barrier to free global trade.

The acceptance and use of ISO standards is very much linked to the neutral status of ISO international standards and their global availability from a single source. The balanced group of experts and the openness and transparency of the standardization process, with public review, promises widely accepted standard content.

A principal goal of ISO is to provide an efficient process to cover all these aspects when writing international standards and to continuously improve the process.

4.4   International Telecommunication Union (ITU)

4.4.1   General

The International Telecommunication Union (ITU) coordinates and promotes international cooperation in telecommunications. It delegates and registers frequencies for sending and receiving radio waves. If there are frequency disturbances in radio communication, the ITU helps to clarify the reason for the disturbances and coordinates activities to rectify the situation.

The ITU supports the development of new technology for telecommunication and coordinates this on an international basis. It provides rules to guarantee system performance and fixes fees. The key issues are given in Box 4.3.

The ITU is an international standardization organization for telecommunication and communication technology. Work on standards is carried out only by electronic means, using Internet connections to all members in companies and organizations. Each member has free access to all standards published by ITU in English only. The financing of the ITU organization is covered by membership fees.

4.4.2   Organization

The ITU is a specialized agency of the United Nations on Information and Communication Technologies founded in 1865 in Geneva, Switzerland. It involves 193 countries in standardization on the basis of a public-private partnership. Beside the member countries, more than 800 private-sector entities and academic institutions are active members of the ITU.

The ITU is an international cooperation between governments of member countries and the private sector and acts as a global forum through which the members work to produce standards by consensus. Today more than 4500 standards are available divided in three ITU sectors – see Table 4.14.

The radiocommunication sector (ITU-R) published reference documents on radiocommunication used by governmental agencies, public and private telecommunication operators, manufacturers, scientific and industrial bodies, international organizations, consultants, universities and technical institutions. These reference documents coordinate the technical use of frequency bandwidths and spectra.

Box 4.3   Key issues for the ITU.

International Standardization of Information and Communication Technologies
All work done on electronic basis
Standards published in the internet to members free of charge
Only English language used
Financing by membership fees

Table 4.14   ITU sectors.

ITU-R	**Radiocommunication**
	Coordination of radio frequencies including satellites
ITU-T	**Telecommunication standardization**
	Operation and interchange of networks
ITU-D	**Telecommunication development**
	Support of public and private organization to develop telecommunication technology

The telecommunciation standardization sector (ITU-T) published recommendations to define how telecommunication networks operate and interact. The more than 3000 recommendations are nonbinding documents. However, these recommendations are used worldwide because of their high quality and their guarantee of interconnectivity of networks and to enable telecommunication services on a worldwide basis.

The telecommunication development sector (ITU-D) supports the technical development of telecommunication technologies and the cooperation with public and private organizations worldwide. The ITU-D gives advice and directs technical assistance to the enhancement of telecommunication technologies. This sector's activities range from policy and regulatory advice, financing of telecommunication, support for low-cost technologies, assistance through human resources management, and the development of rural areas with universal access.

4.5 CENELEC

4.5.1 General

The acronym CENELEC stands for Comité Européen de Normalisation Électrotechnique or in English (as it is commonly used today) European Committee for Electrotechnical Standardization. CENELEC is responsible for standardization in the electrotechnical engineering field and prepares voluntary standards that help to facilitate trade between countries, create new markets, cut compliance costs and support the development of a Single European Market, as stated on their web site www.cenelec.eu (accessed 16 February 2016).

Technical standardization is part of the single European market. Before CENELEC was formed and the EN standards were established, more than 120 000 national standards were valid inside the national borders of the European nations that are now the European Union. After the harmonization process over two decades in the 1970s and 1980s, the number of EN standards in the same area of Europe has been reduced to only 20 000. This is a big gain for the European market and the exchange of goods. This reduced number of EN standards covers the same (or even a larger) technical field. More than 500 million people in the European Union today have the advantages of this market without barriers and fees.

One of the key understandings of CENELEC is that the best way to create markets on an international level, beyond Europe, is to support the IEC as the primary writer of standards. To facilitate this, CENELEC and the IEC established so-called parallel voting on IEC and EN standards under the Dresden Agreement.

CENELEC sees its role in the global market as an organization to support innovative technologies and create competition in the market to make products available industrywide through voluntary standards created by experts from industrial federations and consumer organizations.

The principal requirements of EN standards are consensus among the participating experts, the encouragement of technical development, innovative products, interoperability of products, high levels of safety and the highest possible economically acceptable levels of environmental protection.

CENELEC is a nonprofit technical organization, designated by the European Commission as a European standards organization. It was set up under Belgian law as a result of merging two previous European organizations: CENELCOM and CENEL.

It is one of the officially recognized European Standards Organizations (ESOs), together with CEN, the European Committee for Standards (any standard except electrical standards) and ETSI, the European Telecommunication Standards Institute.

The standardization process for European standards (EN) follows the principles of WTO rules. Many specific terms are used in Europe, which are explained below.

The specific processes involved in mandated standardization, the inquiry procedure, the questionnaire procedure, and the one-step procedure are explained in detail together with the Dresden and Vienna Agreements. Termination, process steps and the weighted voting used in CEN and CENELEC are also explained.

4.5.2   Goals

CENELEC acts as a platform for experts to develop European standards (EN) that facilitate worldwide trade by removing barriers to trade and enhancing economic growth, leading to new markets. There are three goals defined by CENELEC:

- To satisfy the needs of European industry and other stakeholders in the areas of standardization and conformity assessment in the fields of electricity, electronics and associated technologies.
- To lead the improvement of all aspects of product quality, product safety, service quality and service safety in the fields of electricity, electronics and associated technologies, including protection of the environment, accessibility and innovation and to contribute to the welfare of society.
- To support the IEC in achieving the mission: 'To be globally recognized as the provider of standards and conformity assessment and related services needed to facilitate international trade in the fields of electricity, electronics and associated technologies.'

The CENELEC standard-making process is a transparent, consensus-based and open system for all CENELEC members. CENELEC members are committed to implementing the EN standards at their national standards level within a transition period of one year. This creates the single European market, based on harmonized standards. Only a few deviations are allowed at the national level. The CENELEC members with deviations are reminded by the Technical Office to harmonize their regulations and national requirements. This harmonization process makes the European Standards System (ESS) unique in the world.

European standards are a key instrument to support the European economy. The economic benefits of standardization contribute to industrial and societal growth. Standards belong to the knowledge base of the European industry and society.

CENELEC and the NCs work jointly in the interest of European harmonization. Harmonized standards in Europe are requested by the market or by legislation.

The standardization activities of CENELEC increase market potential and technological development, guarantee safety and health to consumers and workers and contribute to a greener world.

4.5.3   Organization

The CENELEC community is an association of members, which are national electrical committees of European countries. In 2015 CENELEC had 33 countries as members and 13 associated members in Eastern Europe, the Balkans, North Africa and Middle East. The member countries are organized in the general assembly (AG) – its structure is shown in Figure 4.6.

The AG has two subdivisions: the technical board (BT) for technical work on standards and the administrative board (CA), led by the presidential committee (PC) at the CEN-CENELEC Management Centre (CCMC) in Brussels.

The AG determines the policy of CENELEC and is composed of representatives of the national committees of the member countries. There is one meeting each year, on invitation of the president.

The CA manages the business of CENELEC and prepares the agenda of the AG meeting. The CA has up to nine members and is appointed by the AG.

The BT controls the standardization process and its speedy execution by the CEN-CENELEC management centres (CCMC) and the officers of the technical committees (TCs). The BT consists of the president and one permanent representative of each member country. The list of responsibilities is long and contains anything that is necessary to manage and operate an efficient standardization process. All matters of organization, working procedures, coordinating and planning standardization works form part of its responsibilities.

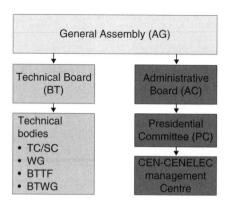

Figure 4.6   Structure of CENELEC.

Box 4.4   CENELEC member countries.

Members:

Austria	France	Luxemburg	Slovenia
Belgium	Germany	Macedonia	Spain
Bulgaria	Greece	Malta	Sweden
Croatia	Hungary	Netherlands	Switzerland
Cyprus	Iceland	Norway	Turkey
Czech Republic	Ireland	Poland	United Kingdom
Denmark	Italy	Portugal	
Estonia	Latvia	Romania	
Finland	Lithuania	Slovakia	

Affiliate members:

Albania
Belarus
Bosnia and Herzegovina
Serbia
Ukraine

The presidential committee is a joint governing body of CEN and CENELEC to coordinate the standardization activities of both organizations.

4.5.3.1   Members

CENELEC is an international organization based on Belgian law. The seat is Brussels and it was founded on 13 December 1972 from two previous organizations – CENEL and CENELCOM. At the time of writing, CENELEC had 33 nation members as shown in Box 4.4. In addition there are six affiliate members. There are more members in CENELEC than in the EU – for example, it includes Switzerland, Turkey, Iceland and Norway.

4.5.4   CENELEC and CEN Specifics

Global Influence

CENELEC and CEN foster the active engagement of European experts in international standardization in the IEC and ISO. Strong links between CENELEC, CEN and IEC/ISO promote coherence between EN and international standards. European standards are key components in trade with countries outside Europe.

Regional Relevance

Common European standards are the basis for a single EU market to strengthen European industry. The European regulatory system, with directives, supports European standards. CENELEC and CEN will drive European standards in cases when there is not international interest. For the EU, EN standards are vital for market harmonization by removing trade barriers for goods and services. The regional relevance of European standards will attract European countries outside the EU to be members of CENELEC or CEN.

Innovation and Growth

European standards support innovation and will encourage market acceptance of innovative solutions. Those who create such standards work closely with researchers and developers in industry. Innovations are brought into the standardization processes of CEN and CENELEC to create innovative products and to provide interoperability and compatibility with new and existing products, services, systems and processes. The CEN and CENELEC communities provide a platform for a long-term environment of cooperation.

Sustainable Systems

CEN and CENELEC ensure the financial sustainability of European standardization through business models that evolve within a fast-changing environment. The costs and responsibility of developing and maintaining standards are borne by the standard makers and users, in order for the system to remain independent and market relevant.

CEN and CENELEC increase the understanding of standardization to encourage the renewal of the expert base and the long-term sustainability of the standardization system (www.cencenelec.eu, accessed 7 February 2016).

Harmonization

The harmonization of national standards of all national members of CEN and CENELEC is one of the key tasks and goals in forming a unique EU market place. This includes setting equal framework conditions for competition in EU markets. The framework for standards in the European Union is given by EU directives that are published by the EU commission in the EU Official Journal. Based on EU directives, the EN standards promote the technical 'state of the art' through design and test requirements. Products and services following EN standards are assumed to fulfil the basic requirements of the EU marketplace. Compliance with EU Directives is a voluntary matter for each manufacturer or supplier of goods and services. Conformity with EU directives is indicated by the CE sign.

Reduction of Trade Barriers

The great success of CEN and CENELEC in recent decades is the reduction of standards applicable in the European Union. The harmonization work of CEN and CENELEC started in 1985 when Europe was ruled by more than 150 000 national standards. The same region, plus some European states not belonging to the European Union today, follows about 20.000 EN standards. This is a reduction of 130 000 conflicting national standards and a great success for the single European market. Figure 4.7 gives a visual impression of this success for about 500 million consumers living in this region.

The trade barriers in 1985 are shown in Figure 4.8 together with the situation today.

New Approach

The harmonization process that started in 1985 in Europe would not have been possible without the support of European industry in cooperation with CEN and CENELEC.

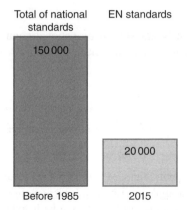

Figure 4.7 Reduction of trade barriers in EU.

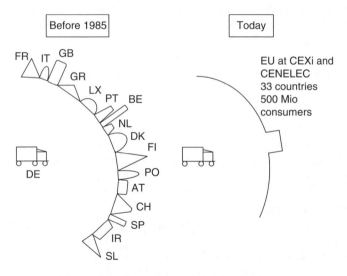

Figure 4.8 Removing trade barriers in Europe.

The Council of the European Union formulated the concept for the technical harmonization of standards in 1985. The European Directives fix general and basic technical requirements in specific fields such as safety, electromagnetic compatibility or environment, to mention some of today's 36 EU directives.

European standards have been worked out by experts in related industries who have been sent and paid for by industry. CEN and CENELEC controlled the open and transparent standardization process. This so-called 'new approach' has proven to be very effective and is market orientated because industry will only send experts if there is market relevance.

A second principle of the new approach is the agreement of all CEN and CENELEC members to transfer EN standards into national standards and to end any of their own standardization work in the same field when EN starts standardization work.

Transfer Obligation

European standards (EN) must be transferred by national committees into national standards within a given timeframe. National standards conflicting with EN standards must be withdrawn.

Standstill Agreement

Once CEN or CENELEC have started to work on a standard, all national committee members must stop their own standardization activities in the same technical field.

What is an EN?

The letters EN stand for 'European Standard' in English and 'Européen Normative' in French. If the standard is applied (and applying EN standards is voluntary) then the requirements of the EN must be followed. It is mandatory for EN standards to be published in English, French and German.

The EN standardization process includes comments by NCs and formal voting among the members of the TC responsible for the standard. When it receives a positive vote, the standard is ratified by the BT of CEN or CENELEC. The publication of EN standards as national standards is then the responsibility of the NC.

National standards in conflict with EN standards are to be withdrawn by the national committee. Every 5 years the NC inquires whether any revision of the standard is required.

Terms in CENELEC

- *Standstill agreement.* As soon as a standardization project has been approved by the BT of CENELEC, work on any other standardization in the same field within the member countries of CENELEC must stop. This standstill agreement is valid until the EN standard is published or, if no consensus is found, has been withdrawn.
- *Unchanged transfer.* The EN standard will be transferred into a national standard without any changes to the document's text and style.
- *Transfer of objective content.* The content of the EN standard will be used but the text and style may be changed.
- *Mandated EN standard.* A mandated EN standard follows the new EU concept of producing the standard under the rules of CEN and CENELEC with experts coming from industry. Standardization work is granted by the European Commission in terms of its mandate. The mandate from European Commission is given to the BT of CEN or CENELEC and from the BT to the related TC. Then the regular standardization process applies and no special marking is used to identify the mandated EN standard. Only a hint in the preface will give this information. In an informative Annex to the EN standard a link is given to the related EU directive. Mandated EN standards must be published in the EU Official Journal, indicating the standard number and title.

Inquiry Procedure and Formal Vote

This procedure allows documents in a public inquiry to be circulated within CENELEC to members of the related TC. The document can originate in CENELEC or in the IEC and comments are collected from all CENELEC member countries. The commenting time frame is 5 months for the first inquiry and 2–4 months for the second inquiry.

The final draft is the next step. This is sent to all member countries for voting. The voting is weighted. The timeframe for the vote is 2 months. If the draft is approved, no reason is required. If there is a negative vote the reason or reasons for the negative vote need to be given with the vote. Editorial comments are acceptable to improve the text and the style of the document. There should be no change to the technical content.

Questionnaire Procedure (QP)

The QP in CENELEC allows the fast transfer of a public document into an EN standard or technical specification. For this reason a primary questionnaire (PQ) is sent to the NCs for comments. If needed, a second questionnaire is sent as an updated questionnaire (UQ) to the NC for comments on the revised document. The timeframe for a questionnaire is 3 months.

At the end of the questionnaire procedure, the final draft will be sent for a formal vote as in the inquiry procedure with effective weighted voting.

Unique Acceptance Procedure (UAP)

The unique acceptance procedure (UAP) is a process in CENELEC that allows it to proceed directly to the voting stage with the proposed document. The proposed document must be accepted by the TC responsible for the related technical field. Acceptance by the BT is linked to an expected positive vote. Usually, this procedure is applied with standards that are used in countries that are national members of CENELEC. The timeframe for the voting is 5 months.

Dresden/Vienna Agreement

The Dresden Agreement for CENELEC and the Vienna Agreement for CEN coordinate the commenting and voting times with IEC standardization. To avoid double work in CENELEC and CEN the comment and voting is in parallel with IEC. This parallel voting increased the publication speed and time to market with the IEC and CEN documents at the same time and national standards of CENELEC members, usually in less than a year.

In most cases the experts are sent from industry to the IEC and in CENELEC no parallel working group is required. Comments come from the NCs. This leads to efficient resource management in CENELEC member countries.

The coordination of the parallel voting procedures in CENELEC and the IEC is continuously monitored by the BT of CENELEC. The content of both documents is in most cases identical.

Types of Standards

There are four principal types of standards in CENELEC available, as shown in Table 4.15.

- The EN is a full standard with mandatory requirements.
- The CEN/TS is a document on the way to a full standard.
- The CEN/TR is a document for publishing technical information with no means to be transferred into a standard.
- A CWA is a CENELEC internal document, which is meant to inform the public.

The detailed processes are explained below.

Table 4.15   Products of CENELEC standards.

EN	European Standard
EN/TS	European Technical Specification
EN/TR	European Technical Report
CWA	CENELEC Workshop Agreement

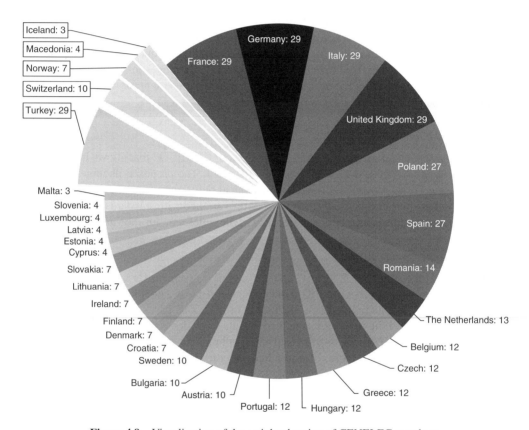

Figure 4.9   Visualization of the weighted voting of CENELEC members.

Weighted Voting

The weighted voting in CENELEC applies to any voting on standard documents EN, EN/TS and EN/TR. The weighted voting system is taken from the Nice Treaty. Figure 4.9 provides a visual overview of the weighted voting of CENELEC members.

The basic criterion to pass a vote in CENELEC is to reach a majority of 71%. Or, seen in the negative way, 29% (a blocking minority of 117 votes) can stop a standardization process. That means that the five biggest member countries can block standards with their 145 votes. To reach a positive result in the voting procedure the 71% requires 288 votes. That means that the 17 biggest national members must approve the standard, which is 50% of the countries. If a majority of all members of CENELEC is not obtained, a second count will take place only taking the votes of the EU member nations into account. That means without Turkey, Switzerland, Norway, Macedonia and Iceland.

The voting will pass if a 71% majority is then reached. On the other hand, the CENELEC member countries which are not part of the EU are not obliged to transfer EN standards into national standards. Usually they do.

4.5.5   Processes

4.5.5.1   General

There are several ways to start the standardization process in CENELEC. Today the most important and most commonly used process is the so-called Dresden Agreement process. In this process CENELEC will accept the IEC work in a parallel voting and commenting process. The Dresden Agreement process is used in more than 90% of the standardization work of CENELEC. This leads to a great accordance with IEC documents.

The normal process for EN standards in the CENELEC process is started within the CENELEC BT, TCs and SCs, in the European Commissions (EU and EFTA) or one of the European technical institutions.

The so-called Vilamoura Process is used in cases when standardization work starts within one or more of the national members of CENELEC. The Vilamoura Process regulates how and when European standards (EN) will have priority treatment before national standardization. It also regulates the standstill agreements of national CENELEC member committees. EN always has priority over national standards. The Vilamoura process is explained in section 4.5.5.4.

4.5.5.2   Organization

The organization of CENELEC is shown in Figure 4.10.

The members in CENELEC are the NCs; they form the AG, which is supported by the CA of CENELEC employees. Technical standardization work is done within TCs. There are 204 TCs today to cover the fields of electrical standardization.

Some TCs have subcommittees (SCs) to split the work. The technical work on standards is done in WGs, which can be initiated by the TC, SC or the secretary report (SR) as an ad hoc working group. The management centres of CEN and CENELEC support the work of the TCs in circulating the documents, voting on documents, evaluating the voting results and watching the CEN and CENELEC standardization processes and rules.

4.5.5.3   EN Standards

EN Standard Procedure

In CENELEC anybody from the member countries has the formal right to propose a new standardization activity. The so called 'anyones right'. The proposal is brought through the NC to the BT of CENELEC for discussion and decision. Other typical routes for new proposals involve them coming from the experts of the NCs and the officers of CENELEC TCs such as chairmen and secretaries. If the BT of CENELEC approves the proposal with a simple majority, the standardization process is started as shown in Figure 4.11.

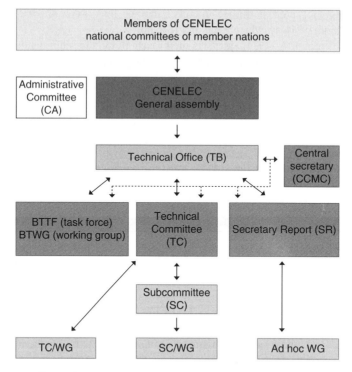

Figure 4.10   Organization of technical work in CENELEC.

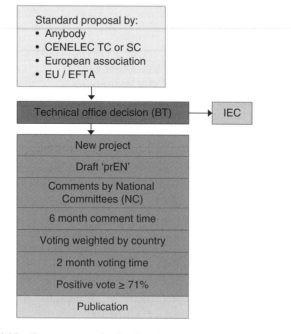

Figure 4.11   European standardization process in CEN and CENELEC.

The CENELEC standardization process can be started from four different sources. The first is related to the right of all member countries. The second source is the TCs and SCs, which have connections with electrotechnical fields through their experts. The third source is related to European electrical technology associations (e. g. ENTSO-E, CAPIEL). The fourth source is so-called mandated standardization by the nations in the EU or EFTA.

The BT of CENELEC decides about the proposal. The first question it considers is whether this work could be done better at an international level with the IEC for global standardization. If CENELEC sees the standard as related to Europe only, or if the IEC is not interested, a new project for standardization is started in CENELEC. At this stage no related national standardization activity is allowed in all member nations of CENELEC (standstill obligation).

A draft 'prEN' will be written by a WG and sent for comments to the NCs for a 6-month period. The revised document then will be sent for voting as a 'final draft prEN' to the NCs of CENELEC. With a need for more than 71% positive votes, the standard will be published as an EN. If the 71% majority is not reached, other ways of publications in CENELEC are checked by the BT, like HD, TS or TR.

4.5.5.4 Vilamoura Process

The Vilamoura process allows a standardization process to be started within the NC of one member state of CENELEC, which prepares a draft standard. This draft standard then will be circulated as a prEN for comments and vote in CENELEC. The Vilamoura process is shown in Figure 4.12.

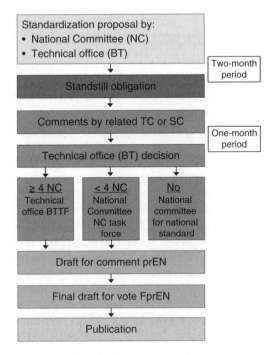

Figure 4.12   Vilamoura process.

One NC proposes new standardization work to the technical office (BT) of CENELEC within 2 months of the start of this activity within the NC. As soon as one other NC of the CENELEC member countries indicates interest in the proposed work, the standstill obligation applies to all CENELEC NCs. Within 1 month, related TCs and SCs in CENELEC can provide comments on the proposal.

Then the BT decides if the proposal will be transferred for international standardization to the IEC or whether to start the work in CENELEC. Depending on the number of interested NCs, three ways of standardization are foreseen. In case of four or more NCs participating in the standardization work, the BT of CENELEC will create a task force (BTTF) or will ask a TC or SC to establish a WG. If less than four NCs are interested, the standardization will be carried out by the proposing NC or a TC or SC will set up a WG. If no other NC is interested the standardization work will be carried out at a national level. In this case the standstill obligation will be withdrawn.

In cases when other NCs are involved in the standardization, the draft prEN then will be commented on by all NCs of CENELEC and the final draft prEN will be voted on according to the weighted voting of CENELEC.

4.5.5.5   Adoption of IEC in EN

In most cases IEC standards are adopted into EN standards by the process outlined in the Dresden Agreement, which regulates the procedure for comments and parallel voting in the IEC and CENELEC.

Where the standard is accepted by the IEC and CENELEC, the National Committees, as members of CENELEC, must transfer the EN standard into a national standard. Deviations in content are only possible in specific cases, for example when national laws or regulations exist, or when particular national conditions (e.g. low temperatures) exist.

In cases outside the Dresden Agreement the way in which IEC standards can be transferred into EN standards is shown in Figure 4.13.

Once the IEC IS document is identified, the CENELEC Technical Office (BT) decides how to proceed. One way is to start a questionnaire to the member countries (NCs) on their opinion regarding transferring the IEC IS into an EN. For the member countries this has the consequence that their NCs must transfer the EN into national standards and withdraw conflicting national standards. In some cases a second questionnaire will be needed to elicit reactions to the comments sent with the first questionnaire and make adaptations to the documents.

In cases when the national members of the technical office do not see any conflict with existing national standards and if the IEC standard is accepted, the so-called unique acceptance process (UAP) can be started. In this case the IEC standard is sent directly to national member committees for voting and commenting. If there is a majority of votes according to the weighted European voting scheme, the new EN will be ratified by the technical offices (BT) of all member nations and published.

Comments from NCs will be accepted. If national or conditional divisions are needed in some CENELEC member countries, they will be marked in the EN standard. The primary goal is to keep IEC and EN content in line and – if changes are wanted – to change the IEC document first.

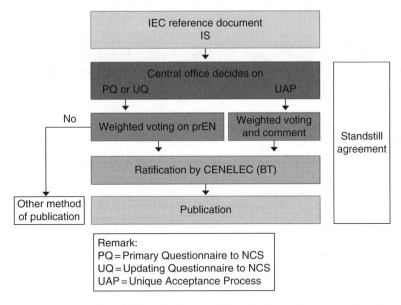

Figure 4.13   Adoption of the IEC standards into EN standards outside the Dresden Agreement.

4.5.6   Transition Periods

European standardization has introduced rules to withdraw national standards of the member nations of CEN and CENELEC. Over a period of time, each country has the obligation to withdraw any national standard in the same technical field as a European EN standard. This process aims to have only one technical EN standard valid in Europe for EU and EFTA countries. There are three dates associated with this rule.

Date of Recognition (dor)

On this date the NC of CEN or CENELEC is informed that the publication of a new EN standard is expected within 1 year. The NC then can use this time to check if any national standard is affected and may identify standards or parts of standards to be withdrawn. Nonconflicting standards or part of standards may remain valid. This is often the case.

Date of Publication (dop)

This date is 1 year after the date of recognition (dor) and indicates the expected publication date of the new EN standard. In case of parallel voting following the Dresden Agreement, the date of publication (dop) is the IEC publication date plus 9 months. The additional 9 months are needed to develop the national translations at least into the mandatory languages of French and German beside English.

Date of Withdrawal (dow)

This date is 36 months after the date of recognition (dor), when for the first time it is indicated that an EN standard will be published. The withdrawal of existing national standards is an important goal of the European Union, as part of its aim to form a unique market without trade barriers. An extension can only be given by CEN or CENELEC if solid reasons for this are provided.

Date of Conformity (doc)

The date of conformity (doc) in an ideal case is the same as the date of withdrawal (dow), because after national standards are withdrawn the national member of CEN or CENELEC are conform to European EN standards.
    The transition periods are shown in Figure 4.14.

Numbering of EN Standards

In case of pure European standards, not taken from IEC, the European standards are EN 50000 numbers. Those standards taken from IEC use the 60000 numbers. If there is no voting majority because there are different opinions over parts of the standard, the document is published as a harmonized document HD with a number. If the standard is only valid in one country the national number, for example DIN-VDE for Germany, is used. See Table 4.16.
    The numbering for draft documents follows the same rule by adding the letter 'pr' before EN ('pr' stands for the French word 'projet').

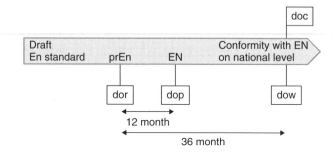

Figure 4.14   Transition periods for European EN standards of CEN and CENELEC.

Table 4.16   Numbering of standards in Europe.

Origin	Number Europe	Number national member (e.g. German)
Pure European	EN 50000:Year	DIN EN 50000: Year or DIN VDE EN 50000: Year
From IEC	EN 60000: Year	DIN EN 60000: Year or DIN VDE EN 60000: Year
Not voted in EN but harmonized	HD XXXX: Year	DIN YYYY HD XXXX: Year or DIN VDE YYYY HD XXXX: Year

National Foreword

Each European EN standard contains a national foreword with the following information:

- national title;
- national foreword;
- national annex, if needed;
- official national language text.

The national title and foreword explain the validity of the standard in the national language. The national annex gives information about national deviations according to the rules.

The official national language text is only mandatory for EN standards in French and German beside the original English version. The English version is taken as authoritative in cases of doubt.

4.6   CEN

4.6.1   General

The European standardization organization has drawn its abbreviation from the French wording of Comité Européen de Normalisation (CEN). Today the English wording is used – European Committee for Standardization – but the original French abbreviation stays. The German wording is Europäisches Komitee für Normung, but German text also uses the abbreviation CEN (www.cen.eu, accessed 7 February 2016).

The CEN is responsible for any kind of standardization in Europe except for electrical standards. The membership of CEN consists of members of the EU plus some members of the European Free Trade Zone (EFTA). The structure of CEN is very close to that of CENELEC. CEN and CENELEC cooperate closely for the harmonization of standards in Europe, with the goal of reducing trade barriers.

It is a private organization founded 1991 under Belgian law. It has 33 national members, 28 EU countries and three EFTA member countries, with one consultant from the European Commission and one from EFTA. There are eight associated members and 17 affiliate members, as shown in Box 4.5.

Box 4.5   Members of CEN.

Members of CEN:

Austria	France	Luxembourg	Slovenia
Belgium	Germany	Macedonia	Spain
Bulgaria	Greece	Malta	Sweden
Croatia	Hungary	Netherlands	Switzerland
Cyprus	Iceland	Norway	Turkey
Czech Republic	Ireland	Poland	United Kingdom
Denmark	Italy	Portugal	
Estonia	Latvia	Romania	
Finland	Lithuania	Slovakia	

One member per country

+---+
| **Box 4.6**   Affiliate members of CEN. |
| |
| **Affiliate members of CEN:** |
| Albania   Georgia   Montenegro |
| Armenia   Israel   Morocco |
| Azerbaijan   Jordan   Serbia |
| Belarus   Lebanon   Tunisia |
| Bosnia and Herzegovina   Libya   Ukraine |
| Egypt   Moldavia |
+---+

Table 4.17   European and non-European partners.

Europe

ANEC	European Association for the Coordination of Consumer Representation in Standardization
AQQA	AQQA Europe
ECOS	European Environmental Citizens Organization for Standardization
ETUI	European Trade Union Institute
ECOMED	European Medical Technology Industry Association
FIEC	European Construction Industry Federation
ORGALIME	European Engineering Industry Association
SBS	Small Business Standards

Non-European

Kazakhstan	Kazakhstan Institute of Standardization and Certification
Mongolia	Mongolian Agency for Standardization and Metrology
Australia	Standards Australia

4.6.2   Members

In addition to the 33 member countries, CEN has 17 affiliates as shown in Box 4.6.

There are eight European standardization bodies as partners and three non-European partners, as shown in Table 4.17.

In CEN more than 60 000 experts carry out standardization work, responsible for more than 11 000 standards.

4.6.3   Organization

The organization of CEN is shown in Figure 4.15.

The organizational structure of CEN is similar to that of CENELEC, which was founded some decades earlier (www.cen.eu, accessed 7 February 2016). All 30 members are represented in the general assembly (AG). The AG forms an administrative board with one vote for each member country. The administrative board is supported by consulting committees (CACCs) for external policy and financial affairs.

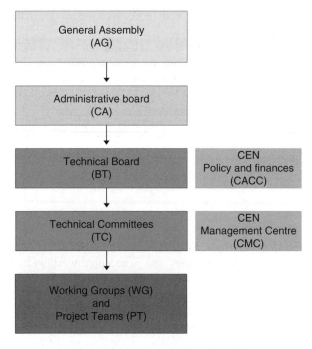

Figure 4.15   Organization of CEN.

The technical standardization work is done by the technical board (BT). It has one representative of each member country with one vote. The BT is the decision maker at any stage of the standardization process.

The technical work is done in technical committees (TCs) representing one field of technology. The division of technical fields is adapted from IEC in so-called mirror committees. If the scope of IEC and CEN is the same then they have the same number. If the scope of CEN varies from IEC then the CEN number has an added 'X'. Each TC has a number of working groups (WGs) or project teams (PTs) to carry out the work on the standards document. In most cases one WG or PT has the responsibility for one standard document. In some cases there is more than one document under the responsibility of one WG or PT.

The CEN management centre (CMC) supports the administrative work of the standardization process by preparing and editing the style of the documents, sending them to national committees for comments and conduct the voting process.

4.6.4   Strategy

The CEN strategic goals are formulated in eight points:

1. To set up a coherent European Standardization System in close cooperation with CENELEC and ETSI.
2. To provide a market-orientated standardization process and find effective ways to bring the standards to the user.

3. To provide standards that are 'state of the art' and which are needed in the market, including the required standardization processes.
4. To provide a stable financing basis for the standardization process including the CMC.
5. To set up good relationships with the Commission of the European Union and EFTAs.
6. To provide European Standards for the Conformity Assessment and the promotion of the European Conformity sign (CE).
7. To check and approve the CEN organizational structure regularly.
8. To be open for partnership and close cooperation with international standardization organizations.

4.7 ETSI

4.7.1 General

The European Telecommunications Standards Institute (ETSI) publishes standards for information and communication technology and is accepted worldwide. It is recognized by the European Union (EU) as a standard-writing organization.

It only works in an electronic way, using the Internet without holding regular working group meetings. The published standards are free of charge to all members of ETSI. It works only in English and it is financed by membership fees.

It was founded in 1988, based on an initiative of the European Commission and its headquarters are in Sophia Antipolis, near Nice in France. In 2015 ETSI had more than 600 members (companies, authorities and organizations) in 62 countries inside and outside of the European Union. See www.etsi.org (accessed 17 February 2016).

The public is connected to ETSI standardization work through national standardization organizations such as DKE in Germany. All voting and comments are made public to members through the Internet for no cost. The type of documents published is the same as in CEN and CENELEC: EN, CEN/TS and CEN/TR.

4.7.2 Members

The European Telecommunications Standards Institute has over 800 members from 64 countries and across five continents. Among the present members are the biggest players in ICT and there are many government and regulatory bodies. But ETSI is the home of small companies, too. Size is unimportant because ETSI members work together as partners in the standardization process.

4.7.3 Organization

The organization of ETSI is illustrated in Figure 4.16.

It has a very lean organizational structure. The general meeting assembles all members and elects a board for the administrative work and the decision-making body. The technical work on standards is coordinated by the TCs, which control WGs, ETSI project groups and partner project groups.

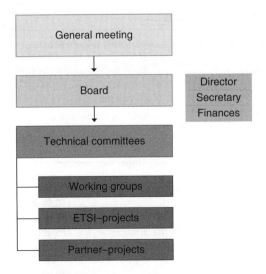

Figure 4.16   Organization of ETSI.

Membership of ETSI started in 1988 with about 100 members. From this time on, ETSI showed a growing number of members with a peak in 2011 of 672 members and about 600 in 2015.

The members come from all European Union and EFTA countries but also the United States, Canada, Brazil, Egypt, South Africa, Russia, Turkey, Iran, India, China, Japan, Malaysia and Australia.

The total number of standard publications also increased since the start in 1988 to several thousand every year. The peak was reached in 2011 with more than 3000 ETSI standards released.

4.8   IEEE

4.8.1   General

The Institute of Electrical and Electronics Engineers (IEEE) is an association to advance technology and innovation for the benefit of humanity. It is the world's largest technical society for professional engineers and covers all aspects of electrical, electronic and computing fields and related areas. See www.ieee.org (accessed 17 February 2017).

The roots of IEEE go back to 1884 when electricity began to have major influence, first with the telegraph and then with electric power and light. The foundation of American Institute of Electrical Engineers (AIEE), the IEEE's predecessor, took place in New York in 1884. The first meeting was held in Philadelphia with such famous companies as Allgemeine Elektrizitätswerke (AEG) (Germany), General Electric (United States), Siemens & Halske (Germany) and Westinghouse (United States) and famous individuals like Thomas Edison, Walther Rathenau, Alexander Graham Bell and Werner von Siemens. Technologies involved at that time were AC induction motors, long-distance AC transmission and large power plants. A secondary focus was on wired communication by telephone and telegraph.

Table 4.18   Overview of IEEE.

Members	>400 000 worldwide
Countries	160
Publications	3 million documents in IEEE explore
Conferences	>1200 per year worldwide
Technical Committees	45 IEEE societies
Standards	>900 active standards

Table 4.19   IEEE global membership.

North America	225 000
Africa	110 000
Europe	57 000
Latin America & Caribbean	18 000
Oceania	10 000
Africa	8000
Total	**428 000**

With the invention of wireless communication, the vacuum tubes and the triode, the Institute of Radio Engineers (IRE) was founded in 1912.

After many years of parallel activities, the AIEE and IRE merged on 1 January 1963 to form the Institute of Electrical and Electronics Engineers with a total membership of 150 000. Since then, the IEEE has grown continuously to more than 400 000 members in 160 countries in 2015. The US-based IEEE of the 1960s had less than 1% of its members outside the United States; today close to 50% of the membership is outside the United States and this is increasing – see Table 4.18.

The IEEE today offers access to more than 3 million technical publications through Xplore, a worldwide search and sourcing tool. More than 1200 conferences are organized by IEEE per year. This makes three IEEE conferences every day around the world.

The global membership of IEEE is heavily weighted towards North America (about 225 000) for historical reasons – the IEEE was founded there. The second largest membership basis is in Asia (about 110 000) but with a strong growth rate in recent years. Europe has about 57 000 members, with strong growth mainly in Eastern Europe.

Smaller membership numbers are found in Latin America and the Caribbean (about 19 000), Oceania (about 10 000) and Africa (about 8000). See Table 4.19.

The IEEE today operates six regional offices. The headquarters is located in New York, with the operation centre in Piscataway in New Jersey, where about 100 staff members work. Offices in Brussels for Europe, Beijing for China, Tokyo for Japan, Delhi for India and in Singapore are small branches with a few staff members. Europe has about 13% of all IEEE members, 1% of the IEEE-SA members and 17% of the corporate members worldwide.

The IEEE is active in publication and organizing conferences in 45 technical communities. Thirty-nine societies provide benefits in specialized technical fields. Through the Standardization Association, the IEEE offers more than 900 active standards (www.ieee.org, accessed 7 February 2016). An overview of the technical societies and councils is given in Box 4.7.

Box 4.7   Overview of IEEE technical societies and councils.

Aerospace and Electronic Systems	Instrumentation and Measurement
Antennas and Propagation	Lasers and Electro-Optics
Broadcast Technology	Magnetics
Circuits and Systems	Microwave Theory and Techniques
Communications	Nanotechnology Council
Components, Packaging and Manufacturing	Nuclear and Plasma Sciences
Technology	
Computer	Power Electronics
Computational Intelligence	Power and Energy
Consumer Electronics	Product Safety Engineering
Control Systems	Professional Communication
Council on Electronic Design Automatization	Reliability
Council on Superconductivity	Robotics and Automatization
Dielectrics and Electrical Insulation	Sensors Council
Education	Signal Processing
Electromagnetic Compatibility	Social Implications of Technology
Electron Devices	Solid-State Circuits
Engineering in Medicine and Biology	Systems Council
Geosciences and Remote Sensing	Systems, Man and Cybernetics
Industrial Electronics	Technology Management Council
Industry Applications	Ultrasonics, Ferroelectrics and Frequency Control
Intelligent Transportation Systems	Vehicular Technology

The IEEE's members elect the leadership, with candidates coming from industry, universities or the public sector from all over the world. An executive director and the staff are employed by the IEEE. The functions of the IEEE organization are shown in Figure 4.17.

4.8.2   IEEE-SA

The IEEE Standards Association (IEEE-SA) is a global organization involved in technical standardization. Today there are more than 900 standards active and 5000 standards are under development. The IEEE-SA has more than 7000 members and 200 corporate members around the world. In standard development worldwide, more than 20 000 experts are active in working groups to do the technical work of standard writing (http://standards.ieee.org/, accessed 8 February 2016). The first IEEE-SA Board meeting was held in June 2013 in Brussels.

There are currently 500 standards in development or under revision. The standardization work is only carried out by volunteers sponsored by industry. The IEEE-SA is an independent organization where participants come together to develop standards independent of any government organization following the rules of IEEE-SA operational procedures.

The European office serves as a link to coordinate the IEEE-SA's activities with CIGRE, ETSI, CENELEC, Eurelectric and other European organizations. The ISO and IEEE-SA have signed a cooperation contract for standardization development to adapt and develop standards jointly. There are currently three technical fields covered: ISO TC 204 for Intelligent

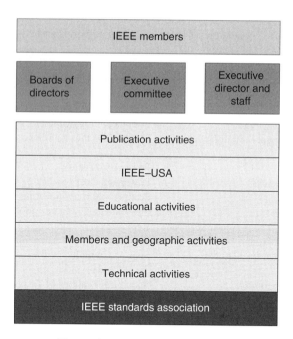

Figure 4.17   -4: IEEE organization.

Transportation, ISO TC 215 for Health Information and ISO/IEC JTC 1 for Information Technology.

The IEEE-SA provides a complete framework for standardization for industry to ensure the rapid introduction of new technologies to the market. It offers the tools to write standards following the WTO requirements for openness, transparency and balance for new documents or revisions of existing documents supported by IEEE-SA staff during any stage of the standardization process.

The IEE-SA supports standard development in two ways. One way is the individual standardization process with individual members of the working groups. The other is based on the corporate programme where entities (e.g. companies, organizations, universities) send representatives to the working groups.

The IEEE-SA's standardization covers a wide spectrum of technology, as shown in Box 4.8.

Two Approaches to Standard Development

In the IEEE there are two ways of writing standards. The first is the individual way and the other is the 'entity' or corporate way. The individual way of writing standards is based on participants who are individuals – people. These individuals represent themselves in the working group or they might be sponsored by a company or an individual to represent their or his interest in the working group. This way of producing standards is necessary to make the work on standards open and transparent to anybody involved. Each individual has one vote when it comes to voting decisions in the standardization work. The voting and commenting is done by ballot, which is set up with a minimum of ten individuals. The ballot group needs to be balanced among the different interest groups, like manufacturers, users, academia and

---

Box 4.8   Broad technology spectrum of IEEE standards.

Aerospace electronics	National electrical safety code
Bioinformatics	Next generation service overlay network
Broadband over power lines	Organic components
Broadcast technology	Portable battery technology
Cognitive radio	Power electronics
Design automation	Power and energy
Electromagnetic compatibility	Radiation/nuclear
Green technology	Reliability
LAN/MAN	Transportation technology
Medical device communications	Test technology
Nanotechnology	

Two ways of standardization development in IEEE

Individual	Entity or corporate
• Participants are individuals, i.e. people	• Participants are entities, i.e. companies, universities, governmental bodies, etc.
• Individuals represent themselves	• Designated representative member and alternative
• One member, one vote	• One entity, one vote
• Individual ballot group of minimum ten individuals	• Project initiation needs minimum of three entities
• Ballot group members are IEEE-SA members	• Entity sends participants to meetings

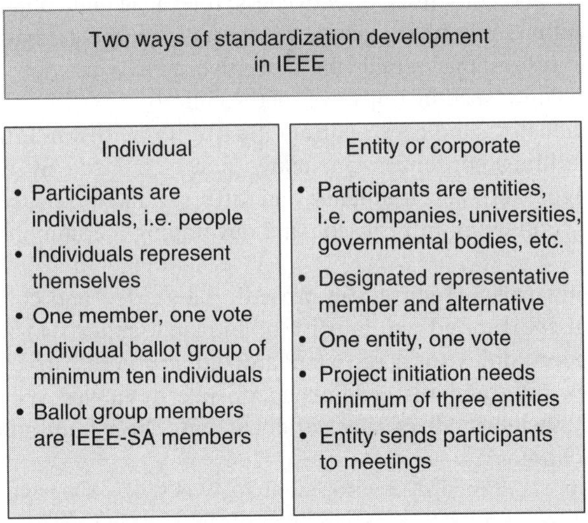

Figure 4.18   Individual and entity standards development.

consultants. To vote and comment on individual standards the ballot group participants need to be IEEE-SA-members – see also Figure 4.18.

The entity or corporate type of standard development is based on membership in working groups of corporate entities – companies, universities, government bodies and so forth. The entity in a working group is represented by a designated working group member (if the representative cannot participate then an alternative representative may do so). When voting on the standard, each entity has one vote independent of how many individuals from that entity are participating in the standardization work. Today there are more than 200 corporate members; about 40% are small or medium-size companies, enterprises, industry associations, academic institutions and government agencies.

To start a standardization project on an entity basis (see Figure 4.18), a minimum of three entities are required to participate actively, send expert members to the working group and pay the appropriate dues for being entity members of the IEEE-SA. This due is only paid once by each entity. The entity standardization approach allows standardization work on a smaller scale as only entities are allowed. This may speed up the standardization process on one hand but might cause problems with global acceptance of the standards.

IEEE-SA Standardization Process

The standardization process in IEEE-SA is open to all interested groups or individuals and is organized in transparent steps. Where conflicts arise there is an internal right of appeal. This consensus-based process is the basis for the widely respected IEEE-SA standards development process, which produces results that reflect the collective, consensus view of participants and enable industry to achieve specific objectives and solutions. The technical work in the WGs in the standardization development examines the various technical solutions and tries to find consensus – see Figure 4.19.

The IEEE adheres to the WTO principles for international standardization: transparency, openness, impartiality and consensus, effectiveness, relevance and coherence. The IEEE-SA standard-development process, which adheres to the WTO technical barriers to trade (TBT) principles, produces results that reflect the collective consensus view of participants and enables industry to achieve specific objectives and solutions.

The IEEE-SA industry connection offers flexibility with an efficient, economical environment for building consensus, incubates new standards by reacting quickly to technological changes and developments and offers related services to prepare new standard projects by collecting information and developing a common view on standardization needs.

Its internal organization is focused on providing a guided and cost-effective process taking into account the limited availability of industry experts in the standardization process. Several opportunities for a fast standardization process are offered in IEEE-SA. There are templates for standards, white papers, peer-reviewed specifications, guides, position papers and other specialized types of document. These hone and refine thinking on rapidly changing technologies.

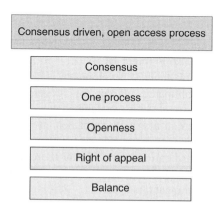

Figure 4.19   IEEE-SA standards process.

Conformity Assessment

The IEEE-SA Conformity Assessment Program (ICAP) bridges the IEEE-SA's standard-development activities with proven testing and certification frameworks. It encompasses all aspects of conformity assessment self-declaration, third-party assessment and testing and interoperability and helps to accelerate the market's adoption of standards for new products and technology.

Conformity assessment plays an important role in the acceptance of standards based on solid test results required by the standard. Each test laboratory must fulfil the technical requirements to carry out valid tests.

Patent Policy

The IEEE-SA patent policy is consistent with the policies of ISO, IEC and ITU. It is designed to balance the needs of those implementing a standard and the rights of intellectual property owners of technology necessary to implement those standards.

The IEEE-SA patent policy contributes to transparency early in the process, confidence in commitments and global applicability.

Patent-related activities with other organizations include a memorandum of understanding with the European Patent Offices (EPO), ongoing discussions with other patent organizations and the European Commission.

Education Programs

The IEEE-SA offers several education programs on standardization. An IEEE-SA standards education web portal is available at https://standards.ieee.org (accessed 17 February 2016) under 'Get involved'.

The IEEE-SA also offers standards education grants to support students and academics for continuous education for practitioners in standardization.

Reaching consensus on technical matters usually does not cause too many problems for the technical experts, because the rules of physics apply to all participants. To avoid conflicts in the working group on how to work it is necessary to offer a due process that is applicable to anybody participating in the working group. The IEEE-SA has set up a set of documents defining due process for standard development in all the steps of the process. One key element of the due process is the need for openness. What is written in the standard document, what has been changed and why it has been changed must be clear to anybody at any time. Therefore, the rules are set and open to anybody. In case of conflict, participants have a right to any participant to an appeal. An appeal group can be established to resolve the conflict.

The management of the standardization development process has to keep the right balance in the process so that it is possible to reach consensus. In an unbalanced process a consensus might not be found. Successful standardization for all participants, manufacturers, users, academia, public or government can only be reached on the basis of a rigorous standards process and its management by IEEE-SA.

International Cooperation

The IEEE-SA started its international cooperation activities around 2005 with contacts with ISO, IEC, ITU at an international level, with CEN and CENELEC in Europe, with the Standards Association of China (SAC) or with Japan (JISC).

The dual logo agreement with the IEC started around 2008. This regulated coordination between the IEC and the IEEE. In cases where no IEC standard is available in a technical field, the IEEE can propose a dual-logo document. The same WG follows both standardization processes in IEC and IEEE in parallel. The final document is then published with two logos: IEC and IEEE.

For example, the standard for *Extended digital interfaces for programmable instrumentation – Part 2: codes, formats, protocols and common commands* is named IEC 60486-2 and IEEE 488.2.

4.8.3   Standards Development Process

In the IEEE-SA, the process of developing standards starts with any interested party in certain technical areas wishing to start work on a new standard, or to revise or extend an existing standard. The first stage, therefore, is to check the existing standardization pool and to establish WGs. The IEEE-SA offers a web site, http://standards.ieee.org/develop/wg/for searching for the scope of existing WGs and http://standards.ieee.org/develop/project/for existing standardization projects and their scope (both accessed 17 February 2016). If there is no existing working group or project available, the IEEE-SA will start the standard development process by identifying the relevant technical society and finding a sponsor in one of the society's TCs. The process for identifying standardization development is shown in Figure 4.20.

Standards Development Process Flow

At the beginning of each standardization process a project-approval process is started to check the consistency of the scope of work, the balanced participation and the overall timeframe. The scope must be clear with regard to what the standard will cover and what is outside the

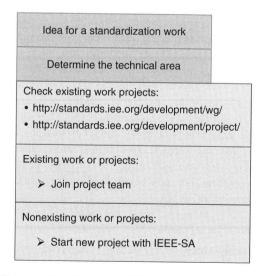

Figure 4.20   Identifying IEEE standards development.

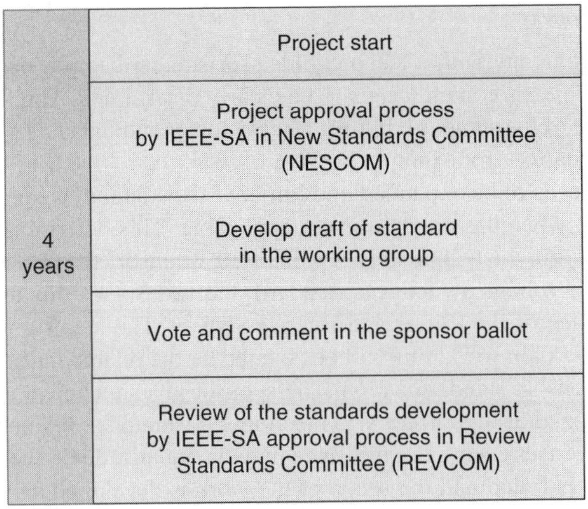

Figure 4.21   IEEE standards development process flow.

standard. There should be no overlap with existing standards and interest groups must be informed about the new work. The experts participating in the work must cover interested parties like manufacturers, users, academia, and authorities, as far as relevant. Representation in the working group must be balanced. The overall time from the start to publishing the standard must not exceed 4 years. Within this timeframe and after project approval the WG will have several draft documents before formal voting on the document in the balloting process. At the end, before publication, the IEEE-SA Standards Board will approve the document. Then the standard is published – see Figure 4.21.

The project-approval process is organized by the project authorization request (PAR) form, which needs to be sent for approval to the IEEE-SA New Standards Committee (NESCOM) at the IEEE-SA Board. Once the PAR is approved, the WG can start to write and circulate drafts. After the WG decides that the content of the standard is ready for approval the document will be sent into the sponsor ballot process. This process is coordinated by IEEE-SA to set up a balloting group of any interested group balanced of manufacturers, users, academia, authorities, government and any other interested body. The balloting group then will comment and vote on the document and the working group will have to address all comments and revise the document accordingly.

Before the revised document is published by IEEE the Review Committee (REVCOM) of the IEEE-SA Board will check if the rules for writing standards in IEEE have been followed and any comments have been addressed. The goal of the IEEE-SA is to have the work done in 4 years. This is accomplished in most cases. If more time is required the working group will have to ask the IEEE-SA for a PAR extension and get approval by the IEEE-SA Board. In a period of 5 to a maximum of 10 years, the WG responsible will decide to start a revision process with a PAR or to extend the validity of the standard by another 5 years. Standards that are out of time may be withdrawn and shifted to the archives to be stored with no revision ongoing.

Project Approval Process

Any individual or company with a technical idea for standardization may start a project. In the first place the project's content needs to be clearly determined. The first task is to define the headline and title of the standard. It must be easy to understand and be clearly distinguishable from other standards – for example *Devices of more than 1000 V.* In the text of the scope the technical field needs to be explained and limits of the standard scope are to be specified. It must also be clear when the standard is not applicable. This determination is best done in close cooperation within the technical group that will 'sponsor' the project.

If you take a high-voltage switchgear standard, the sponsor of this project would be the Switchgear Committee of the Power and Energy Society. The sponsor does not financially support the standardization work; this will be covered by the related industry, which will send experts. Industry sponsors standardization. The sponsor gives technical guidance in the subject and offers organization support for WG meetings and annual or biannual committee meetings. In some urgent cases even more meetings may be organized per year by the sponsor.

Once the sponsor is found and the scope of the work is developed and agreed, the project will be sent for approval to the NESCOM of the IEEE-SA Board. The basis for the process is the PAR. On a quarter-year basis IEEE-SA will check, comment and approve the PAR. See Figure 4.22.

The standards of the IEEE are written by working groups. The basis of the work may be existing draft documents or specifications. In rare cases the work starts from scratch. The working group needs to follow IEEE-SA policies and procedures defined by the technical committees. The draft documents are circulated to the working group members at each stage of development, being reviewed and revised until the working group decides that the work has been done. The next step of the document then is the balloting process.

The Project Approval Process is shown in an overview in Figure 4.23.

After a sponsor is identified and supports the project, the PAR will be completed and submitted to the NESCOM. This will check whether the PAR is complete, ask for changes by the working group and make recommendations to the IEEE-SA Standards Board. The Standards Board needs formal approval of the PAR before the work can start in the working group. The PAR is a legal document to obtain permission to work on a standard under the authorization of the IEEE-SA Standards Board. In this case the IEEE extends the umbrella of indemnification to people working on an authorized standards project if they follow the rules given by IEEE-SA.

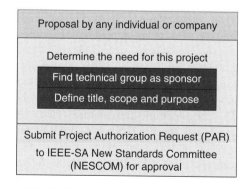

Figure 4.22   IEEE standards development approval process.

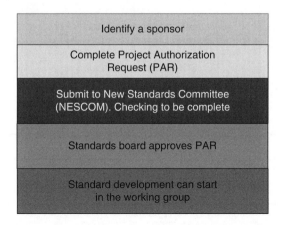

Figure 4.23   Overview of project approval process.

Balloting Process

To allow any interested person or company to participate in standard development the so-called balloting group is established by the IEEE-SA and the chairman of the WG. It is the responsibility of IEEE-SA to ensure the openness and balance of the balloting. All members of IEEE-SA can be a member of the balloting process. They need to join IEEE-SA Ballot and pay an annual fee. The web-based balloting process tool is then open for all standardization activities of IEEE-SA. As a member of the balloting team he can make comments and vote on the standard (http://development.standards.ieee.org/my-site, accessed 8 February 2016).

The vote is to 'approve' or 'disapprove' the standard. Where a vote is to 'disapprove', reasons must be given. Votes to disapprove need to be addressed by the working group on the basis of technical arguments. Only if all negative voting has been addressed can the standard be published. To address negative votes does not mean that the WG must agree with the proposed changes linked to negative votes but they need to have some technical reasons why they will not adapt the proposed changes.

The ballotters, in most cases, approve the draft document in the ballot process with comments to be taken into account by the working group.

The standard approval process has four steps. In the first step the final draft document of the working group is sent to the REVCOM for reviewing the standards development process. The REVCOM recommends that the IEEE-SA Standards Board to approve the standard. The IEEE-SA Standards Board approves the draft standard and then the standard goes through final editing and style checking for publication, as shown in Figure 4.24.

The sponsor balloting process is shown in Figure 4.25.

The first step of the balloting process is to form the balloting group. An invitation to ballot is sent to all IEEE-SA members and they can join by checking a Web-based tool. The balloting group is split into interest categories, such as producers, users, general interest and authorities. None of these basic groups can have more than 33% of all votes. Once the balloting begins the balloting group will be closed and remains static for the whole balloting process. Membership of the IEEE-SA or payment of a fee is required to participate in the ballot. With the closing date of the balloting it is necessary to have a minimum of 75% return from the balloting group and 75% of the ballots (votes) have to be positive for the vote to pass.

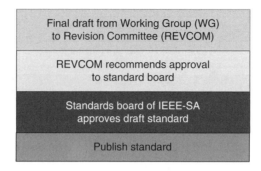

Figure 4.24   IEEE approval process for publication.

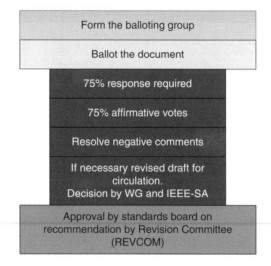

Figure 4.25   Sponsor balloting.

There is an attempt to address negative ballots but there is no mandate to resolve all ballots. When consensus is reached at 75% approval a timeline is established for the resolution of the comments.

In case of large changes on the document based on the comments a recirculation of the revised draft and comments is recommended to obtain consensus in the WG and balloting group.

After the document has been revised the document will be prepared for the REVCOM. This must show evidence of attempts to address the comments.

Revision Committee Approval

The final approval of a standard before publication is made by the REVCOM as shown in Figure 4.26.

The final check by the REVCOM is to ensure that the standard has been written according to the IEEE-SA rules. The REVCOM will check all the documents from the PAR, the draft,

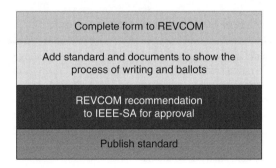

Figure 4.26 Revision committee (REVCOM) approval.

the comments, the ballot and the responses to the comments received with the ballot. The REVCOM meets every 3 months together with the IEEE-SA Board.

4.8.4 Power and Energy Society

The IEEE Power and Energy Society (PES) is one of 45 technical societies in IEEE organizing worldwide conferences and technical publications and has responsibility for more than 350 standards. The PES has worldwide activities with the main basis in the United States. See www.ieee.org (accessed 17 February 2016).

The membership of the PES has been growing since 2010 when the topic of electric energy supply came under increasing public scrutiny and when new technologies like renewable energy and intelligent network control and management were entering the scene. The membership of PES grew from about 20 000 in 2008 to more than 32 000 in 2014. These members are linked to 213 chapters, an increase of 12, and 65 student branches, an increase of 29. About 60% of the PES members is located in North America, 14% in Europe, 12% in Asia and 10% in South America.

The technical activities are split into 18 TCs and six coordinating committees. They offer a wide network with the brightest minds in power and energy technology.

The PES offers several technical meetings each throughout the world. In North America there is a joint technical committee meeting in January, a general meeting in July and, every second year, the transmission and distribution conference and exhibition in April.

In Europe, Asia and South America, PES offers the Power Technology Conferences on general technological aspects of Power and Energy and the Integrated Smart Grid Technology (ISGT) Conference on the digitally controlled and managed electrical supply of the future. Conferences on region-specific topics are also organized.

The policy of PES is to offer industry the possibility for the exchange of knowledge, the discussion of trends and making contacts between experts. The competence of PES in the power and energy field provides the basis for organizing successful conferences.

Another stronghold of PES is the large basis of more than 350 technical standards. This is more than 40% of all IEEE standards. These standards are in responsibility of 18 TCs. The committees are listed in Box 4.9. These committees have subcommittees and WGs for standard writing and to conduct technical tutorials.

Box 4.9   Technical committees of PES.

---

1. Intelligent Grid and Emergency Technologies Coordinating Committee
2. Marine System Coordinating Committee
3. Conductors
4. Electric Machinery
5. Energy Development
6. Energy Storage and Stationary Batteries
7. Nuclear Power Engineering
8. Power System Analysis Methods
9. Power System Communications and Cyber Security
10. Power System Dynamic Performance
11. Power System Operations, Planning and Economics
12. Smart Buildings, Loads and Customer Systems
13. System Protection
14. Substations
15. Surge Protective Devices
16. Switchgear
17. Transformers
18. Transmission and Distribution

Box 4.10   Transactions of the IEEE PES.

---

1. Power Systems
2. Power Distribution
3. Energy Conversion
4. Sustainable Energy
5. Smart Grid

The PES has a close relationship with CIGRE and with IEC. Cooperation agreements are also made with Japan (IEEJ/PES) and China (CSEE).

IEEE offers several ways to publish technical papers. Power and Energy Society publications can be made in the form of conference papers, which are made public by the conference proceedings. A different way to publish papers in IEEE PES is through the different transactions (see Box 4.10). Transactions are published monthly or in a several-month-long sequence. All publications from IEEE conferences or in IEEE transactions are published in IEEE Xplore and are publicly accessible through the Internet. IEEE Xplore is the largest source of technical publications in the electrotechnical field.

4.9 International Coordination

4.9.1 General

Globalized business requires increasing coordination of standards. Many products and solutions today are designed for the world market. Large international companies but also an ever increasing number of small and medium-size companies require international standards for a global use. Market pressures and the limited number of experts available for technical standardization are leading to more coordination of international standardization organizations. The IEC, ISO, CEN, CENELEC, IEEE and many others have established coordination agreements. In the following some important agreements for the electrical industry are explained.

4.9.2   IEC and CENELEC

4.9.2.1   Dresden Agreement

The coordination of IEC and CENELEC is organized through the so-called 'Dresden Agreement'. In principle, IEC and CENELEC agreed that the voting process and with this the processing of comments on draft standards are coordinated so that they are timed to run in parallel so that comments from both sides can be treated on the same document. The IEC document has the lead and changes are made here first to keep the content of IEC and EN documents the same. See Figure 4.27.

When the standardization project in the IEC is started on basis of a NP, at the IEC level the work is carried out within the IEC WG. Committee drafts (CDs) are circulated only in the IEC. When the IEC releases the next step, the CDV, the Dresden Agreement says that this document is circulated for vote and comments in IEC and CENELEC at the same time (parallel votes).

Comments are collected in IEC and CENELEC and processed in IEC. The votes are evaluated separately in the IEC and CENELEC and only if both are positive can the next step in the parallel process be taken.

The next step in the IEC is the final draft international standard (FDIS) and this IEC document is then circulated in the IEC and, at the same time, in CENELEC as the final draft preliminary European norm (FprEN). Only if a positive vote is reached on the document

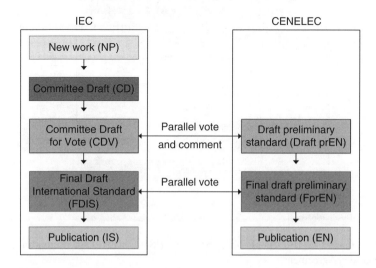

Figure 4.27   IEC-CENELEC parallel voting (Dresden Agreement).

according to IEC and separate CENELEC voting rules can the document be published as an International Standard (IS) in IEC or as a European Norm (EN) in CENELEC.

This process is mandatory for IEC International Standards (IS) in CENELEC. For other IEC publication types, like technical specification (TS), technical reports (TR) or as IEC/IEEE dual logos, parallel voting according to the Dresden Agreement is voluntary for CENELEC. A formal decision of the related technical committee of CENELEC is required.

4.9.2.2 Outside the Dresden Agreement

The decision regarding whether an IEC standard will be transferred to CENELEC as an EN standard is taken by the BT. The voting members of the BT are the NCs. The process is shown in Figure 4.28.

Once the decision is made by the BT of CENELEC, no other European member state is allowed to start standardization activities in the same technical field until the CENELEC process is finalized in a positive or negative way.

If CENELEC votes positively, the EN is published and the European member states of CENELEC must transfer the EN into their national standards. If the vote is negative the member states are allowed to start their own standardization activities in the same field.

There are two ways in principle for CENELEC to transfer IEC standards to EN. One way is to ask the NCs for their vote and for comments and in the second step for a final vote to publish the EN. The other way is a one-step vote where only the final vote is requested with the possibility of the NCs giving comments on the document. The decision about the route that will be chosen is taken by the BT of CENELEC. The one-step vote process is faster but risks failure if no weighted majority is achieved.

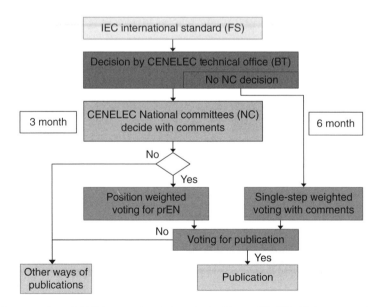

Figure 4.28 Transfer of IEC International Standard (IS) to CENELEC European Standard (EN) outside the Dresden Agreement.

In case of a negative vote, the BT will check whether other forms of publication in CENELEC can be chosen in the expectation of obtaining a majority in the vote. Such other publications might be, for example, a harmonized document (HD), a technical specification (TS) or a technical report (TR).

4.9.3   ISO and CEN

4.9.3.1   Vienna Agreement

The Vienna Agreement coordinates the standard processes in each organization to obtain parallel voting as in the Dresden Agreement between the IEC and CEN. The Vienna Agreement was signed in 1991 and was established with the goal of avoiding the duplication of work in the ISO and CEN and of increasing the transparency of the work in ISO and CEN and the member countries. Standardization work under the Vienna Agreement recognizes international requirements by the ISO and regional requirements by CEN in a common commenting and discussion process.

To coordinate and monitor the Vienna Agreement, a Joint Co-ordination Group (JCG) of the ISO Technical Management Board (ISO/TMB) and the CEN Technical Board (CEN/BT) has been established. The JCG monitors the reports on specific issues of concern, existing mechanisms and on proposals for improvement.

5

National Organizations

5.1 General

The national organizations for international and national standardization work are in most cases reflecting the international organizations IEC, ISO and ITU.

The technical work is then reflected in so called mirror committees with the same responsibility of scope as on international level. The larger the industry of a nation the larger is the national committee on standardization. Nations like USA, Germany and Japan do reflect almost any IEC, IS and ITU technical work on their national level.

In the following some national organizations are explained in their role of international standardization.

5.2 Germany

5.2.1 General

The importance of exports for German industry has given Germany a widespread international business orientation. This is true for large companies like Siemens but also for small and medium-size companies. They use international standards for their international outreach in business even without a global presence and local offices. With international business, international standards are needed (see www.din.de, accessed 18 February 2016).

The German standardization organizations, the DIN and the DKE, are actively participating in standardization at the European and international level. At the European level, the primary standardization organizations are CEN for anything except electricity and CENELEC for electrical standardization. At the international level, the preferred partners for German standardization organizations are ISO for anything except electrical standards and IEC for electrical standards.

Practical Guide to International Standardization for Electrical Engineers: Impact on Smart Grid and e-Mobility Markets, First Edition. Hermann J. Koch.
© 2016 John Wiley & Sons, Ltd. Published 2016 by John Wiley & Sons, Ltd.

Table 5.1   Overview of international standardization cooperation of Germany.

	National	Europe	International
General	DIN	CEN	ISO
Electrical	DKE	CENELEC	IEC
Telecommunication	DKE	ETSI	ITU

For the telecommunication sector, at the European level, ETSI is the partner for Germany. At the international level it is the ITU. An overview is given in Table 5.1.

In Germany, in principle, international participation in standardization activities is voluntary for all interested parties. This means that who will participate in IEC, ISO, CEN and CENELEC is decided on a case-by-case basis. There is no requirement by law or any authority to participate.

Participation in any international standardization activity is transparent to all interested parties through the discussions and decisions in the related technical committees and the publication in the *DIN-Mitteilungen*. Any interested person or entity can obtain information about the international committees in which Germany will participate and what the basis for the decision is.

When the German technical committee is deciding whether or not to participate in international standardization, it considers the question of market relevance. Does the proposed standardization work have any relevance to the market? If the decision is no, then Germany will not participate. This happens very seldom.

The basis for the decision on participating in international standardization is the consensus of the participating technical committees in Germany. This is sometimes a discussion between different interests like industry, science or society and any other interested party. The process is open to all interested parties.

Once a consensus is found at the German level the goal is to bring this consensus to a worldwide acceptance in the international standardization.

In many cases Germany is a driver in proposing new standardization activities whether for a new standard or to start revision of an existing standard. This driving role comes from innovations in industry and leadership in some technical areas.

In the competitive field of technical innovation it is important for Germany to start standardization projects at an early stage. This allows it to influence the content of a standard and give direction to technical development. In many cases other national committees with similar innovative industries act in the same way and a certain level of consensus can be found during the early stages of the standardization process.

As a consequence of Germany's international orientation, national standards (EN) are in most cases equal in content to European and international (IEC and ISO) standards. For this reason Germany is a permanent member in the IEC, ISO, CEN and CENELEC and the work of their technical committees is often reflected in its own national committees. CEN and CENELEC standardization work is coordinated with IEC and ISO standardization work through the Dresden and Vienna Agreements. It is mandatory to transfer CEN and CENELEC standards into national standards. There is no requirement for ISO and IEC standards to be transferred into national standards.

Germany's active participation in international standardization is reflected in the number of active German experts. The DKE nominated 1200 experts to the IEC and 550 to CENELEC.

Table 5.2 Development of international, European and German standardization.

	1980	1988	1998	2008	2011
International	6000	10 000	16 000	23 800	25 500
European	490	1120	9100	19 000	19 300
German	19 000	20 500	26 000	31 000	33 050

This includes 40 leading positions as chairman or secretary in CENELEC and about 70 leading positions in IEC in the electrotechnical field. This is 20–30% of the whole IEC or CENELEC leadership. A similar picture can be found in ISO and CEN.

The development of international European and German standardization activities based on DKE publications (https://www.dke.de/de/Seiten/Startseite.aspx, accessed 2 March, 2016) is shown in Table 5.2.

The numbers in Table 5.2 show a strong increase in standards over the last four decades. In Germany, there has been an increase of about 70% but this is low compared to the increase of 400% with the IEC and 4000% in Europe. These figures reflect the impact of globalization and the increasing importance of international standardization. And there is no end in sight.

In Germany, standardization in the electrotechnical field has a long history and starts with industrialization. One of the first standards published worldwide is a little booklet from 1896 on high-voltage safety (*Sicherheitsregeln für elektrische Hochspannungsanlagen*). In 23 pages the total spectrum of high-voltage equipment like switchgear, cables, motors, transformers and insulators is explained and requirements are specified. For insulators, it stated that wood cannot be used to insulate because it absorbs water and it will lose the insulation capabilities of dry wood. For solid insulated cables it is specified that a high-voltage cable must at least work without flash over of the insulation for more than 30 minutes after it is flooded with water. The little document is the basis of IEC 61936, which provides installation rules for high-voltage substations. Besides taking a national view of standards, from the beginning Germany has had a global view on electrical equipment and its standardization. This can be found in German standardization organizations DIN, DKE and VDE today and is reflected in their wide engagement in standardization.

5.2.2 DKE

5.2.2.1 History

The Deutsche Kommission Elektrotechnik Elektronik Informationstechnik (DKE) is the German National Committee, representing Germany in international standardization organizations like IEC, ISO, CEN, CENELEC and the IEEE. The German understanding of standardization includes any regulation related to safety for the technical fields of electric, electronic and information technology.

The basis of DKE is a contract between VDE and DIN on the standardization process in Germany and representing Germany internationally (see Figure 5.1). This contract was signed on 13 October 1970 and work started on 1 January 1971. The basis of this contract is the so-called *Normenvertrag* between DIN and the German government signed on 6 June 1975 (https://www.dke.de/de/Seiten/Startseite.aspx, accessed 2 March 2016).

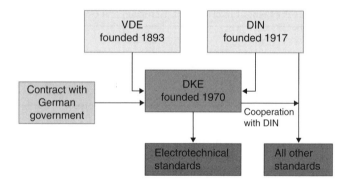

Figure 5.1 DKE for German standards.

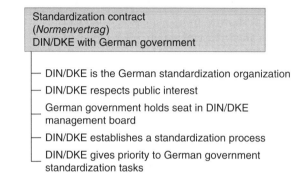

Figure 5.2 Standardization contract DIN/DKE and German government.

The *Normenvertrag* (Figure 5.2) specifies that the DIN/DKE, as the national standardization organization, is responsible for Germany and represents Germany in nongovernmental international standardization organizations. DIN/DKE looks after the public interest in standardization and sets up a process through which the public can contribute to standardization.

The German government has a seat on the management board (*Lenkungskreis*) of the DIN and DKE and can send representatives of governmental organizations in working groups when relevant. The DIN/VDE has set up rules on how to write standards and has a quality-management system in place to control the process in each working group. The DIN/DKE will prioritize standardization work coming from the German government.

The DKE board has representatives from interested national organizations with a total of 27 seats as shown in Table 5.3.

5.2.2.2 Goals of the DKE

The goals of DKE are to work out accepted technical rules and to publish them as DIN standards. If they are related to safety matters the standard will be classified as a DIN/VDE standard. Technical rules will be prepared before the standardization process is started and the results are made public. The DKE participates actively in European and international specifications for standardization (https://www.dke.de/de/Seiten/Startseite.aspx, accessed 2 March 2016).

Table 5.3 Seats of national organizations in DKE.

Organization	Function	Seats
ARD/ZDF	Production- and Technology Commission ARD/ZDF (Public TV)	1
BDEW	Bundesverband der Energie- und Wasserwirtschaft (Energy and Waters)	1
BG ETE	Berufsgenossenschaft Energy, Textil, Elektro e.V.	2
BM	Bundesministerien (Wirtschaft, Arbeit und Telekommunikation) (Government Ministries)	3
BM V	Bundesministerium (Verbraucherschutz) (Government Ministries Consumers)	1
DB	Deutsche Bahn AG (Railroad)	1
DIN	Deutsches Institut für Normung e.V. (Standards)	2
GDV	Gesamtverband der deutschen Versicherungswirtschaft e.V. (Insurances)	1
VDE	Verband der Elektrotechnik Elektronik Informationstechnik e.V. (Electrical Institute)	2
VDTÜV	Verband der technischen Überwachungsvereine e.V. (Test institute)	1
VGB	(Technische) Vereinigung der Großkraftwerksbetreiber (Large Scale Power Plants)	1
VIK	Verband der Industriellen Energie- und Kraftwirtschaft e.V.	3
ZVEH	Zentralverband der deutschen Elektro- und Informationstechnischen Handwerke (Electrical Workers)	1
ZVEI	Zentralverband Elektrotechnik- und Elektronikindustrie e.V. (Electrical Industry)	7

The targets of DKE are related to safety, compatibility, market orientation, consensus, German interest, quality and conformity. The safety aspect is a basic view of DKE in standardization to protect people and equipment. This view includes testing of equipment and human protection in the environment of professional work. The compatibility of components within technical systems is an important requirement and contributes to competition in the market and economic system solutions. This aspect of market orientation supports the rapid market success of new technologies through technical support with information presented in standards and specifications. Reaching consensus in the technical work of all interested parties and providing the knowledge needed for all participants is an important goal of DKE in organizing the working groups.

As an organization, the DKE sees its main role as offering a place for dialogue, fostering competence and engagement in standardization. The promotion of dialogue between all interested parties will strengthen standardization and will keep participants engaged. With this engagement, participants extend their personal knowledge and competences. The results are high-quality standards offering a high safety level in all related devices, equipment, systems and services in the electrical field.

The DKE aims to publish accepted, market-orientated and technical rules in its standards. It supports the progress of technical development and produces positive changes. This is reflected in the daily work of its employees.

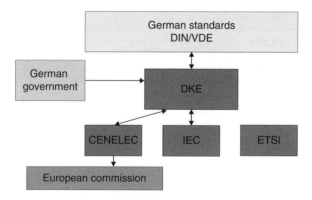

Figure 5.3   Workflow of the DKE.

The goals of the DKE are focused on supporting Germany's position as an international, export-orientated economy. Standardization is seen as a strategic support instrument to develop markets for the benefit of business and society. It reduces the need for regulations and government intervention in business and reduces trade barriers. The standards support technical convergence when new developments enter the market. The efficiency of standardization work is itself a focus of the DKE, which offers processes and tools for the experts involved in standardization.

The workflow in Figure 5.3 gives an overview of how the DKE is working between national and international organizations. The DKE publishes national standards in German, with a DIN/VDE number. It also nominates experts to CENELEC, IEC and the ETSI.

On the political side, the German government cooperates with the DKE directly as member of the management board and it also has influence on the DKE through the European level of standardization.

5.2.2.3   Structure of DKE

The DKE is a private entity, operated by a management board and an operative manager supported by about 100 employees located in Frankfurt / Main, Germany. The technical work is divided into nine technical areas as shown in Figure 5.4.

Germany is well represented at the international level in the IEC and at the European level in CENELEC. In the IEC, in 2014, Germany had 1250 experts nominated and in CENELEC it had 550 experts active. With 17 chairmen and 24 secretaries in CENELEC and 36 chairmen and 34 secretaries in IEC, Germany also plays a leading role in these organizations and their working bodies. Germany also works actively in the technical field through working groups (WGs) and maintenance teams – see Figure 5.5.

The experts come from industry and are supported financially by their companies. They are only engaged in some parallel work with the IEC and CENELEC, such as work on graphical symbols or vocabulary. The process of standardization is supported by the DKE by providing information and guidance to experts from industry in IEC or CENELEC working groups.

Working groups in the DKE are limited to a maximum of 15 members to keep the groups effective and so that they are able to reach a consensus in technical discussions. Where there is greater interest, subgroup representatives are sent to the DKE to represent fields of interest.

In 2014 the DKE had 3500 experts with a total of 290 committees and subcommittees, including more than 40 WGs.

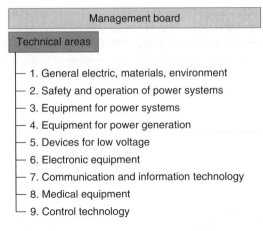

Figure 5.4 Structure of DKE in principle.

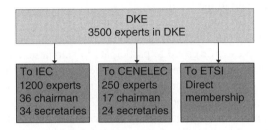

Figure 5.5 DKE activities.

Organizing Standardization

Since 2008, the DKE has established a new sector responsible for organizing standardization. Beside the technical areas there is a service unit inside the DKE to improve the internal and external processes and integrate innovative areas and new issues into the standardization process.

Potential rules are defined and published for continuous analysis of requirements for standards. The DKE is continuously searching in Europe and internationally to identify topics for future standardization work.

The following technical topics have been identified and supported: solar technology, optical data transmission, ocean wave technology, nanotechnology, electro mobility and telemonitoring in medicine.

In close cooperation with universities, the DKE sponsors pilot projects on standardization with the science to standards (STS) program.

A research program has been started on ambient assisted living (AAL) with a group for interoperability and quality criteria. The publication *Interoperability of Ambient Assisted Living – System Components* was in preparation at the time of writing.

Another topic in preparation is electromobility and electric energy. Here the technical fields of interaction and interoperability between the wide use of electromobility and the risks to the electric power supply are investigated. This includes the technical possibilities of contact-free charging for electric vehicles. For this, the EU founded the project for Ambient Assisted Living Platforms and the German government funded projects for living in cooperation with hospitals.

The principle of industry supporting the experts in standardization covers more than 80% of the costs related to standardization. It also covers participation in international standardization, including travelling costs. The remaining 20% of the cost is related to the activities of DKE. About 95% of the income is from selling the standards to the user. There are no membership fees and no governmental financial support.

Today, 80% of German standards are based on IEC standards; about 15% are based on European standards and only 5% are pure German standards. In 1990 the situation was different: 40% were German based and 40% were European based and only 20% were IEC based.

5.2.3 DIN

5.2.3.1 General

The Deutsches Institut für Normung e. V. (DIN) is a private entity founded under German law. By contract with the German government, it is responsible for standardization in Germany and for representing Germany in European and international standardization, except for electric standards, which are related to the DKE.

The DIN coordinates economic, research, consumer, public sector, worker associations, trade and standardization from the proposal to the publication of the standard. In principle it allows every interested person or group to participate in standardization and acts as a round table for all interested groups. It reflects a global view of German industry. All interested parties in Germany have access to standardization, regardless of their size. International standards are transferred into national standards by the DIN.

The standards of the DIN are free of conflict and have a unique form. An electronic infrastructure is provided with Livelink. This connects the DIN, ISO and CEN in one platform.

5.2.3.2 Structure of DIN

The DIN has more than 1000 members, who form a management board (*Präsidium*), which elects a CEO and its support functions. The standardization is related to three subdivisions as shown in Figure 5.6.

Figure 5.6 Structure of DIN.

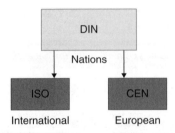

Figure 5.7   DIN in Europe and internationally.

The largest sector in the DIN is related to mechanical engineering, metal and services, which make up more than 50% of the membership. More than 75% of the members belong to small and medium-size companies with less than 1000 employees. Of the 60 non-German members, 70% are from Switzerland and Austria and the rest are spread all over the world. The DIN represents Germany in CEN at the European level and in ISO at the international level (see Figure 5.7).

The DIN has 400 employees. Most are located at its headquarters in Berlin. There are about 30 000 experts coordinated by DIN, which are supported by the industry. In 2014 there were more than 30 000 standards in place and published. Each year 2000 new or revised standards were added or replaced.

5.2.3.3   Financing

The core financing for the DIN comes from selling standards. More than 70% of the financing comes from selling standards, 15% from industry-sponsored projects and 10% from public funding. Only 5% of the income comes from membership fees.

The transfer of standardization work over the last 30 years went from national to international. In 1986, 90% of the standards were national; in 2014, 80% of the standards are international. The transfer to international standardization had the consequence that the work in the DIN also went international.

The DIN cooperates worldwide with countries on all continents except North America and Australia. It is active in fields such as microbiology, nanotechnology, optical technology, biotechnology, information and communication technology, new materials and energy efficiency.

Standardization also has limitations when it comes to regulations and laws. Market access is not only guaranteed by standards; regulations and laws also play an important role and can be a strong barrier against market access. The influence of national rules is difficult. On an international basis, the WTO is politically active in supporting free trade.

The European Union (EU) releases European Directives through the EU. These European Directives must be transferred to national laws by the member states of the EU. The Directives set framework conditions for national standards. Examples are the EMC Directive, the Environmental Directive, the Low Voltage Directive and the Machinery Directive.

The standards committee (*Normenausschuss*) coordinates the standardization activities. The work is then split into the three technical fields shown in Figure 5.8 where committees (*Arbeitsausschuss – AA*) and subcommittees (*Unterausschuss – UA*) share the work.

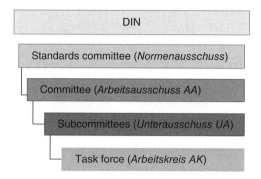

Figure 5.8   Principal steps to write a standard in DIN.

5.2.3.4   Standardization Process

About 90% of German standards are from the IEC and ISO at the international level or CEN and CENELEC at European level. The remaining 10% of standards have to follow the process shown in Figure 5.9.

The German standardization process is, in principle, a two-stage process. At the first stage, a preliminary document is created and printed on pink paper (*Rosadruck*). The second stage is the published version, printed on white paper (*Weißdruck*). This is an historical process that makes it clear to every reader whether a document has been finalized.

After a proposal for a new standard is made, the national standardization technical committee of the related technical field will decide if the work will be started or not. The technical committees are structured in technical fields (*Fachbereiche*). The standardization process shown in Figure 5.9 is organized as follows.

The proposal for a new standard can come from anybody and will be evaluated by the DIN/DKE together with the related technical committee. When approved, a working group will be established and the project can start. A timeframe is set for the working group and working drafts are circulated within the working group. After a consensus on the content of the standard is reached by the working group, a preliminary standard will be circulated by DIN/DKE on pink paper (*Rosadruck*) to all interested parties with a date given for comments. The timeframe for comments is 3 to 6 months. All comments are collected and all commenters are invited by the DIN/DKE for the discussion of possible changes on the document. After the discussion is finished a revised preliminary standard is prepared by the working group and sent to DIN/DKE for final editing. The final version of the standard is then produced by the DIN/DKE in the style of DIN standards. The publication will be certificated by the chairman of the technical committee of the DIN/DKE and printed on white paper (*Weißdruck*).

Coordination with International Standards

Coordination with international standards is in the responsibility of the DKE for electrical standards and the DIN for all other standards. International coordination of electrical standards is linked to IEC and for Europe to CENELEC. The coordination of IEEE standards with German standards is made through the dual logo agreement of IEEE with IEC. Pure IEEE standards are not coordinated.

The coordination between IEC and CENELEC is organized by the Dresden Agreement. Coordination between the ISO and CEN is organized by the Vienna Agreement. For IEC and

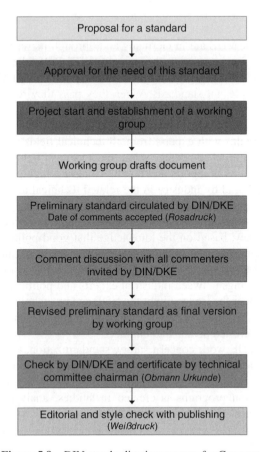

Figure 5.9   DIN standardization process for Germany.

CENELEC the German national committee collects the national comments and presents them to IEC and CENELEC.

The voting and commenting processes for the committee draft of vote (CDV) in the IEC and the draft preliminary European standard (Draft prEN) in CENELEC are in parallel (Dresden Agreement). The numbering of IEC and CENELEC is the same. After the European standard is published as an EN, a national standard is mandatory.

5.2.4   VDE

5.2.4.1   General

The Verband der Elektrotechnik Elektronik Informationstechnik (VDE) has 36 000 members including 1300 enterprises, 8000 students and 6000 young engineers. It is organized into different societies related to technical fields. It operates as an independent test institute, offers certifications and prints and publishes books and standards.

The VDE was founded in 1893 by the initiative of Werner von Siemens and other leaders of electrotechnical companies in Germany. The headquarters of VDE is in Frankfurt/Main, with branch offices in Berlin and Brussels.

5.2.4.2   Goals

The main activities of the VDE are in Germany and, through its representation, in Brussels, also in Europe. It is active worldwide through its 'VDE-Prüfstelle' – the test institute in Offenbach, close to Frankfurt. Here all kinds of electrical products, mainly consumer products, are tested according to relevant standards. When they pass they receive the VDE test sign (VDE-Prüfzeichen).

One big goal of the VDE is to maintain a dialogue with the public and politicians by organizing conferences and talks with experts from all technical fields of the electrical industry. The VDE has therefore split its competences into five technical fields. These topic-orientated discussions and publications in reports and brochures have a high public and political value because the results are found by industry in the related technical area. The VDE reports are seen as sources of independent information.

In this respect, the VDE aims to be an independent, reliable and experienced consultant for policy makers in Germany, based on the knowledge that good political decisions and good technical information can be made by independent experts. The VDE can organize independent expert groups and lead the cooperation of experts with political institutions – in particular, regular information meetings between industrial experts and political institutions and policy makers are organized under the leadership of the VDE and by technical experts from industry. In the same way, the VDE offers conferences on information technology development and tutorials to teach any interested person and engineer.

The main topics of VDE work concern safety, standardization, technology development, information for professional engineers, the young engineers programme, research programs and innovation support. Sponsored by VDE and the industry basic information for political information and education programs is created in studies, analysis, position papers and initiatives on special topics.

5.2.4.3   Structure of the VDE

The structure is shown in Figure 5.10. The VDE has 36 000 members, either with personal or corporate membership. The leadership is elected from delegates who, in turn, are elected from all members. The 'Präsidium' is the leading body.

The work is done in five technical sectors (*Fachgesellschaften*): ETG for energy technology, ITG for information technology, DG BMT for biomedicine technology, GMM for microelectronics and GMA for automation and control. The testing and certification of VDE is organized within a separate business unit to publish books, standards and conference papers.

ITG

The Informationstechnische Gesellschaft im VDE (ITG) was founded in 1954 and had 10 200 members in 2014. It covers topics related to the information society with focus projects, services, applications, TV, film industry, electronic media, audio-communication, communication technology, high-frequency technology, microelectronics and nanoelectronics.

ETG

The Energietechnische Gesellschaft im VDE (ETG) was founded in 1974 and had 10 700 members in 2014. The ETG covers topics related to electrical power including the electric

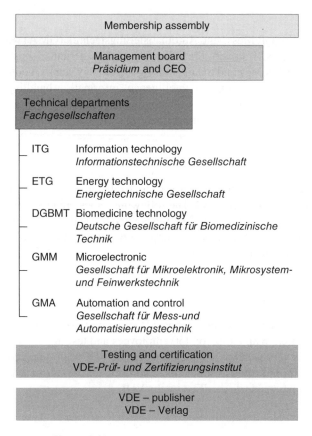

Figure 5.10   Structure of VDE in principle.

power supply, generation, transmission, distribution and electric trading and marketing. Other areas are applications of electric machines, trains and electric vehicles, system integration, materials, insulating systems, diagnostic methods, behaviour of contacts, switches and circuit breakers.

DG BMT

The Deutsche Gesellschaft für Biomedizinische Technik im VDE (DG BMT) was founded in 1971 and had 1700 members in 2014. It is active in the following technical areas: communication and information technologies, graphical systems, diagnostic systems, therapy systems, medical technologies, biotechnologies, applied clinic engineering.

GMM

The VDE/VDI Gesellschaft Mikroelektronik, Mikrosysteme und Feinwirktechnik (GMM) was founded in 1984 and had 8200 members in 2014. It covers the following technical fields: microtechnology and nanotechnology for manufacturing and applications, mechanic-electric-technology, design of printed board technology, electronic design automation, and electromagnetic compatibility.

GMA

The VDE/VDI Gesellschaft Mess- und Automatisierungstechnik (GMA) was founded in 1973 and has 13 000 members. It covers the following technical fields: sensors and measuring systems for process technology and manufacturing technology, actors, controllers, robotic and electric-mechanical systems, computer intelligence, information handling, communication, visual technologies, automation systems including engineering, operation, diagnosis and training, application of control and automation systems.

FNN

The Forum Netztechnik/Betriebssysteme im VDE is an expert forum founded in 2008 to exchange knowledge and experiences on safe and reliable design and operation of power transmission and distribution networks. The following topics are covered: technical network connection requirements, technical safety management, quality of electric power supply, measuring and metering of electric power supply, protection and control of power networks and geoinformation systems.

5.2.4.4   VDE Test Laboratory

The VDE operates a test laboratory in Offenbach, Germany, and certifies products and services as a third-party certifier. With a total of 450 employees and testing facilities for all consumer products, the VDE is very well known in Germany. The VDE sign stands for tested safety and high product quality as specified in standards. Two-thirds of the Germans know the VDE sign and only buy consumer products certified by the VDE.

5.2.4.5   VDE Publisher

The VDE is one of the largest German publishers in the electrical field with more than 2500 German standards and more than 500 book titles and professional journals in the technical areas shown in Table 5.4.

5.2.4.6   Young Members

Young engineers and students at the end of their study time can participate in VDE activities. Today more than 12 000 students and young professionals are using this offer for networking and job exchange. There is no membership fee required to use this service. Special activities like 'Invent a Chip', 'Youth Research', 'Think Engineers' and 'FOCUS' are offered at public schools on a yearly basis with attractive awards and presentations of results at VDE conferences.

Table 5.4   VDE professional journals.

Technical field	Journal
Electrical engineering and control	*Electronik und Automation (ETZ)* (German)
Communication	*Innovationen der Kommunikationstechnik (UTZ)* (German)

5.2.5   Norm Contract

In Germany, in 1975, the standardization process was given a new contract to coordinate and optimize the efficiency of standardization work. The government of Germany and the DIN reached an agreement that assigned the DIN and DKE to represent Germany in international standardization and to organize standardization activities in Germany for nongovernmental standardization. The DKE was founded in 1970 by the DIN and the VDE for standardization activities.

In the contract, the DIN and DKE are obliged to organize the standardization work in a way that recognizes the public interest and the public has access to the standardization work at any time. There should be a high level of transparency. The German government has voting seats on the DIN and DKE boards and receives priority service when it requires standardization. Government representatives might participate in standardization work if this is needed or required. The DIN and DKE have set up bylaws with DIN 820 to define the standardization process.

All safety-related standards in Germany are the responsibility of the VDE, whereas all other standards are the responsibility of the DIN (see Figure 5.11). Any standard concerning electricity is regarded as a safety-related standard.

The goals of DKE are fixed in the Norm Contract as shown in Box 5.1.

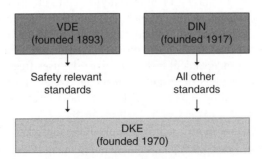

Figure 5.11   Founding of the DKE.

Box 5.1   Goals of DKE in standardization.

Safety
Compatibility
Market orientation
Consensus orientation
Representative for any interested party
Quality
Conformity

Safety

Safety is the key requirement of standards in the view of the DKE. This goes back to the very beginning of electrical industrial development in the middle of the nineteenth century. At that time it was clear to the innovators and market leaders like Edison or Siemens that the key for the success of this new technology was the safety of users. Electricity cannot be seen, smelled, or sensed in any way by humans but when touched the danger is immediate. If people are afraid for their life they will not buy electrical products. So business success was very strong related to safety aspects.

This influenced the standardization work from the beginning and still does so today. Electrical products and systems must be built so that they are safe at any time for workers, users and the public.

Compatibility

Standards need to provide compatibility between devices and systems. This compatibility should allow users to operate equipment by different manufacturers so that they are not dependent on only one manufacturer. This prevents monopolistic situations and increases competition in the market.

In many cases technical products offered in the market are complex and it is not possible for potential users to evaluate the features of a product. For example, in the case of a circuit breaker, the manufacturer states that the equipment can switch a certain current, for example a 40 000 A short-circuit interruption if there is a short-circuit fault in the network. The related standard gives design and testing rules for such features and independent testing laboratories can test the equipment and give a test protocol to confirm the required switching capability.

With this certificate, the manufacturer can prove that the device is functioning correctly and that the user can trust the independent test. This gives the user freedom to choose between several equipment manufacturers of circuit breakers and creates a market for such devices. Standards are the key to creating compatibility between devices and systems.

Market Orientation

The market orientation of standards is an important requirement if standards are to be accepted in the market. Standards must create a market and serve a wide range of possible users. Without market orientation, standardization work would become bureaucratic and would lack any benefit for the market. It would waste the efforts of experts, build up costs that would not be paid back by business and would be without benefit to the public.

Rapid technical development can only be converted into market success with a standardization policy that reflects the need to open the market for new, innovative products. Standards and technical specifications are the tools to create market-orientated standardization work. For this reason, before a standardization project is started it is necessary to check its market relevance.

Consensus Orientation

To reach consensus on technical matters it is necessary to open the standardization process to any interested party in the related technical field. The available knowledge needs to be brought into the technical discussion and agreement by the experts is required. Controversial positions of interested parties need to be discussed and a common position needs to be found. Only if a consensus is possible can a standard be published.

Representation of Interest

The German view of standardization is to develop international and European standards in the first place. The reduction of trade barriers is the key goal – a worldwide opening of markets for the benefit of the public. National standardization shall be avoided as much as possible and is only the last choice for standardization work in Germany.

Quality

Standards need to describe the state of the art in technology. They should be of a high quality level. This means that the standards must follow up-to-date technical rules, must be consistent with other standards and must be widely accepted in the market by the user of the standards. The standards must be orientated on what the market needs.

Conformity

The conformity assessment rules must be applied to the evaluation of test results of requirements defined in standards. Conformity assessment plays an important role in the worldwide acceptance of standards.

5.2.6 Standards and Laws

The role of standards in Germany covers an important link between products and systems at the manufacturer's level and governmental responsibility for product safety covered in laws and regulations. The principle is shown in Figure 5.12.

The production of devices and systems and their operation is organized within companies following private rules and manufacturing and operational standards. Private rules are not public and are the basis of specific knowhow and technical expertise. Based on this knowledge, standards are used by industry for operating devices and systems manufactured by different companies to open a technical market on the basis of competition. These standards are public and any interested entity or person can participate in the work on standards. They create publicly accepted rules based on industrial consensus. Standards and product rules offer technical flexibility regarding how to solve technical requirements in the most economical way.

Laws and regulations outline what is accepted in terms of safety risks, care for the environment and the protection of society. These laws and regulations are publicly accessible and known to all players in the market. They do not give requirements on technical matters, where it is open to industry to find the best solutions. The requirements are general and do not specify details. This is important as it avoids limiting innovation. Laws and regulations should

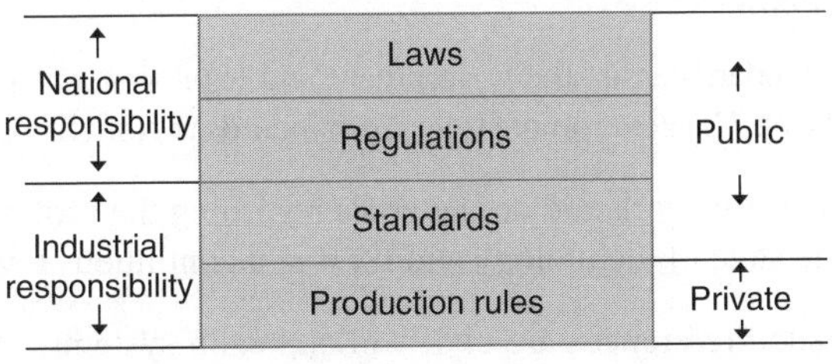

Figure 5.12 Principle of the relationship between standards and laws in Germany.

not require particular detailed technical solutions because it takes too long to change laws and this would be a barrier for technical innovations for the benefit of the society. Legislation needs to focus on safety and environmental requirements.

Legal Position of Standards

Standards are seen as legal basis for evaluating the state of the art of technology and to investigate whether a product follows an accepted rule for the technical device or a system following two principles:

- An accepted rule for technology is a technical explanation by experts that reflects the status of technology accepted by a representative majority of experts in this field. This is represented by a standard.
- The state of the art of a technology represents the status of technical possibilities at the time when the product or system was designed. That means that products, processes and services are based on knowledge generated by science, technology and experience.

Germany follows these four principles:

- standards are applied voluntarily;
- standards are mandatory only when required in a contract or when a law requires a standard to be applied;
- standards are definite, accepted rules and offer security when applied in contracts;
- in case of an appeal standards are seen by the justice as 'first evidence'.

Standards are applicable in different areas of law:

- public law: safety, environment, health;
- civil law: care;
- criminal law: danger, protection.

Science and technical rules are used in law in the following three ways:

- Incorporation – complete or partial use of standards in laws.
- Referencing – references to standards used outside a law:
  ◦ as a fixed reference to a standard with a given publication date;
  ◦ as an open reference to a standard in its latest published version;
  ◦ as an extended reference to European regulation with the safety requirement and its national standard for detailed requirements.
- Use of general legal expressions as 'accepted technical rules'.

Standards and Deregulation

In Germany, the DIN standards assist the government and legal system to reduce the number of laws and the details in laws. The government refers to standards to fulfil basic requirements in laws:

- Laws create the legal framework and define goals regarding the protection of society.
- Standards specify the state of technology and fix it in a continuous revision process.

Typical areas with such references are civil works, health systems, environmental protection and machinery. For example, on 10 November 1971 a crane collapsed on the last segment of a bridge crossing the River Rhine. The bridge segment and workers fell 30 m down into the

Rhine and 13 workers were killed. Two standards, DIN 4114-1 and DIN 4114-2 for the stability of cranes, gave relevant rules and requirements. The justice stopped the legal case against the design engineers and the construction control team because the requirements of DIN 4114 had been followed. The reason for the accident was an unknown load. In the opinion of independent experts, the standard represented the accepted state of technology at the time. The accident could not have been be avoided. The required revision of the standard included the unknown load and was published as DIN 18800. This standard was then incorporated into the law for civil construction works.

Standards as State of Technology

This example shows the importance of the 'state of technology' in the German legal system. German judicial practice evaluates the 'state of technology' in terms of the standardization process according to the DIN/DKE rules.

In each case the justice requires verification of the rules given in a standard from independent experts. It should confirm that the standard is correct in explaining the related technology. For this it is required that the standard should cover the complete technical field and be up to date and the standardization process should be relevant to the case. The basis for this standardization process is given in DIN 820.

Regulations to Enter the Market

In Germany several special laws and product safety rules regulate the market access. Some examples are given in the following lists:

Special Laws

Special laws include laws covering:

- food and devices;
- medicine;
- civil work products;
- electromagnetic compatibility;
- chemicals.

Product Safety Laws

Product safety laws include laws covering:

- electrical products;
- toys;
- machinery;
- printers;
- elevators.

Criminal Responsibilities

In Germany, criminal responsibilities are outlined in two paragraphs of the criminal law (*Strafgesetzbuch*):

§ 222 StGB – Careless Death
Whoever causes the death of someone through carelessness will be sentenced to up to 5 years of prison or a monetary penalty.

and

§ 229 StGB – Careless Injury
Whoever causes injury to someone through carelessness will be sentenced to up to three years of
prison or a monetary penalty.

5.2.7 Standards Process

In Germany the standardization process follows the WTO but shows some specific differences
compared to the IEC, for example. One important difference is that the DIN/DKE plays a more
active role and the position of the national organization is strongly integrated into the German
standardization process. The standardization process follows six steps as shown in Figure 5.13.

The public proposal starts the standardization process with a standards proposal
(*Normenantrag*). This proposal is then checked by the related technical committee and this
technical committee initiates a standard template for the work in the related working group.
The working group will prepare a manuscript for a standard draft (*Manuskript für Norm-
Entwurf*) usually in several steps within the working group of the technical committee until it
is ready for public comments (*öffentliche Stellungnahme*), which are treated by the technical
committee under the lead of the inspection authority (*Normenprüfstelle*) to prepare the manuscript for standards (*Manuskript für Norm*) including all comments received by the public.
Then the inspection authority will finally check the manuscript for standards before publication
as a DIN standard. In case of a conflict of interest regarding the standard's content, an arbitration
process will clarify the situation before publication as a DIN standard. This principal process
of standardization in Germany is split into 23 steps as explained below in Table 5.5.

Anybody in Germany can initiate a standardization project by sending a standard proposal
to DIN or DKE. The related technical committee will then be asked to express their opinion
of the proposal. In some cases the proposal will be covered by an existing working group
with the next revision of an existing standard. In some cases the proposal will be refused with
solid technical argumentation and in some cases the proposal leads to a new standard project.
If it is decided to proceed with a new project, the DIN or DKE staff will start the process for

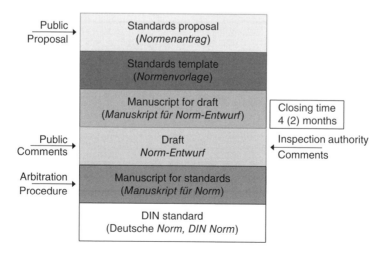

Figure 5.13   Standardization process in Germany.

Table 5.5 Sequences of standardization process in Germany.

	Step	Responsible
1	Standard proposal	Anybody (*Jedermann*)
2	Standard demand determination	DIN/DKE Technical Committee (*Fachkreise mit Geschäftstabelle*)
3	Standard project start	DIN/DKE staff (*DIN/DKE Geschäftsstelle*)
4	Standard project public indication	DIN/DKE Project Control (*DIN/DKE Referat*)
5	Consultation technical committee	Working Group
6	Discharge standard	(*Arbeitsgremium DKE*)
7	Manuscript for draft standard	Project Control DKE (*Referat*)
8	Internal precheck by DKE	Project Control DKE
9	Send manuscript to DIN	(*Referat*)
10	Internal precheck by DIN	Project Control DIN
11	Standard draft with 2- or 4-month comment period	Publisher (*Verlag*)
12	Comments to DIN or DKE	Anybody (*Jedermann*)
13	Comment consulting (invite all commenters)	Project Control (*Referat und Arbeitsgremium*)
14	Revised standards manuscript for standard	Working Group (*Arbeitsgremium*)
15	Final manuscript for standard text for public information	Project Control DKE (*Referat DKE*)
16	Check file and send to DIN	Project Control DKE (*Referat DKE*)
17	Provide certificate of technical committee chairman	Project Control and chairman DKE (*Referat und Obmann DKE*)
18	Printing, registration, check and correction	Project Control DIN (*DIN Normenprüfstelle*)
19	Monitor-print and final draft approval	Project Control DKE (*Referat DKE*)
20	Send final draft to DIN	Project Control DKE (*Referat DKE*)
21	Final check and printing clearance	Project Control DIN (*DIN Normenprüfstelle*)
22	Print standard	Publisher (Beuth-Verlag)
23	Deposit copy for technical committee	Project Control DKE (*Referat DKE*)

standardization to set up a working group, nominate experts and give a timeframe for the complete standardization process.

In step 4 the public will be informed of the new work by the journal *DIN-Mitteilung* – a monthly publication that indicates the status of each standardization project in Germany. This information is published by the staff of DIN and DKE with technical support from the technical committee and related working group.

The real work on the standard document is done in step 5 with several rounds of consultation in the working group of the technical committee. Here the document is written and further developed, step by step. This sequence may last 1 or 2 years. Once the working group has reached the opinion that the work is done it will discharge the standard and DIN or DKE will prepare the manuscript of the draft standard. After a check by the project control of the DKE in step 9 the manuscript will be sent to the DIN. An internal check by DIN project control is the last step before the standard draft is published with a period of 2 or 4 months for comments. *DIN-Mitteilungsblatt* also makes the publication of the draft standard public.

In step 12 anybody is invited to comment on the draft standard within the time period and send the comments to the DIN or DKE. All commenters are invited by the project control to discuss and clarify all comments. A revised standard manuscript will be prepared by the working group and the final standard manuscript will be sent from the DKE to the DIN. In the DKE, the certificate of the chairman of the related technical committee will be provided and signed by the chairman (*Obmann*). In the DIN the project control (*DIN-Normenprüfstelle*) will check the data file of the standard and will prepare the print, registration and, if necessary, corrections – but with no change in the technical content.

Before this is distributed to the public a monitor print will be released by DIN and sent to the DKE for a final check. After clearance is given in step 21, the publisher, Beuth Verlag, will print the standard and provide an electronic version. The technical committee will receive a copy of the standard.

5.2.8   Parallel IEC-CENELEC Process

For all electrical standards the international level of IEC and the European level of CENELEC are closely linked to DKE processes. The idea behind this is that, for an export-orientated nation like Germany, standardization must, in the first place, take a global view to reach global acceptance of products and solutions. In second place, Germany looks to Europe where more than 50% of German products are sold and where France is still the country with which it has the largest level of trade in goods, before the United States.

Thus the leading standardization organization for Germany is the IEC followed by EN standards. The DKE cooperates closely in the standardization processes of the IEC and CENELEC as shown in Figure 5.14.

The international standardization work for Germany is coordinated in contracts between IEC, ISO, CENELEC and CEN in the following agreements:

* Dresden Agreement (IEC–CENELEC);
* Vienna Agreement (ISO–CEN).

Comments on IEC, ISO, CEN or CENELEC documents for Germany are only presented through the DKE and DIN. The IEC and CENELEC processes for electrical standards, under the Dresden Agreement, are coordinated in a parallel process with comments and voting in the same timeframe. The same parallel process is coordinated in the Vienna Agreement for all other standards beside electrical ones. Published EN standards are required to be translated into the German language and published through the DKE or DIN.

The standardization process in more than 90% of all electrical standards in Germany starts with the IEC and then the cooperation process in Figure 5.14 is followed. When new work (NP) is indicated, Germany will comment through its national committee, the DKE. The next step in the process is to send comments from the DKE to the IEC for each Committee Draft (CD). The CD is also used in Germany to start the German Public Comment Process. This is a requirement of the German standardization process rules that any standard coming from IEC and CENELEC to be published in Germany must go through the public comment process so that the public, informed by the *DIN-Mitteilungsblatt* (www.din-mitteilungen.de, accessed 7 February 2016), has the chance to comment on the standardization proposal. This is very actively used and comments collected are transferred by the DKE to the IEC committee drafts, CD and CDV, and to the CENELEC draft, prEN. At the final draft level (FDIS) in the IEC and in CENELEC at prEN, the German national committee votes on the document through DKE.

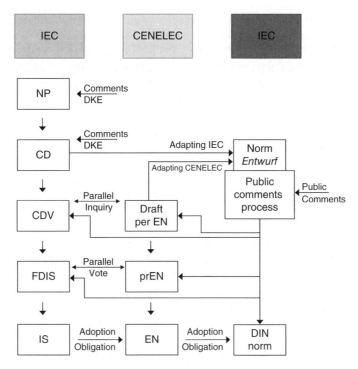

Figure 5.14   Cooperation between IEC-CENELEC-DKE from a German view.

The IEC standard (IS) is transferred to CENELEC in the parallel voting process without any change in the content. The common changes of IEC and CENELEC are made on the FDIS document in the IEC only. The EN standard is then adopted by Germany.

5.2.9   Organization of DIN/DKE Standards

The standardization work of the DKE and DIN is organized in an autonomous way. That means that the business management decides how its money will be spent. They are independent. The standardization work follows the ten principal rules shown in Chapter 2. These rules are fixed in the bylaws of DKE and DIN, including the rules for the work of the experts in the working groups. The rules for how to assembly a working group to reflect all interest groups are also defined in the bylaws.

The DKE and DIN see themselves as a round table for manufacturers, trades, users, workers, services, science, technical inspection companies, the state, which means anybody interested in standardization. For this the group of experts has to examine the state of technology and to take into account new knowledge.

The experts in the DKE and the DIN bring together the required expertise, offer a balance of interests in the group, publish and revise the standard documents and ensure that the standardization process is followed according to the rules and that the public is invited to participate by making comments and giving informal input to the standardization group.

To keep the groups of experts of a workable size, the DKE and DIN require them not to exceed the maximum number of 15 experts in one technical committee. If more experts are interested, the status of a guest is provided and groups can be split into subgroups of interests with one delegate representing each subgroup on the technical committee.

All the rules are given in DIN 820 and the VDE bylaws are given in VDE 0022. The rules for the experts' work in the DKE and DIN follow the general principles explained in Chapter 2.

The DKE and DIN are very strict in applying the rules to set up a balanced group of experts that is authorized by their associations, public authorities or science institutes like universities to carry out the technical work in the working group. There is no direct membership of manufacturer or user companies, only the delegation through its association. Each technical field of interest needs to be adequate represented in the working group, which is mostly important for safety-related standards. All results of work in the working group must be published to the public in time to obtain feedback and comments from all interested persons or entities. The timeframe for this work is fixed in the DIN rules and follows the IEC and ISO (international rules) and EN (European rules). All comments and any feedback will be reflected in the standard revision process by the experts of the related working group. If there is disagreement, an arbitration process can be started as explained in the DIN 820 rules.

The most important reason for the DKE and DIN to be trusted is the strict rules defined in DIN 820. The principal rules outlined in Chapter 3 are also applied by the DKE and DIN. Clause 7.1 of DIN 820 states that 'Anybody may start the process of standardization.' There are no preconditions or exclusions. The process is completely open. Anyone interested in the standard development process may contribute. The standard must not unduly advantage single individuals. To control this in the standardization process is the most important work of DKE and DIN. The need to follow the general rules of compliance in business and not to allow single and individual interests to prevail is the core value of any standardization organization.

Clause 9.1 of DIN 820 states that the content of a standard shall be found by consensus among participating individuals and groups, with participating parties attempting to find a common accepted position in the working group. This should be possible without the need for a formal vote in the working group. This rule is based on the conviction that, on technical questions that follow the principles of physics, it should be possible for all participating experts to find a common position on each technical topic. Formal voting at working-group level with experts participating always has some political content, which is not well suited for finding a good technical solution for a standard. To reach a good solution the experts in the working group need deep discussions and sometimes a little more time.

The main goal of any standardization work is to find consensus, which is defined in DIN EN 45020. It is common agreement with the absence of any opposition to essential parts of the content of the standard. It is characterized by a process that attempts to bring the views of all parties together and to clear all counterarguments. Consensus does not mean unanimity.

In the real practical world the experts are searching for a text formulation that is acceptable to everybody in the working group. In most cases this text can be found by wording the phrases in such a way that nobody feels that his interest will be harmed by the text. The wording should cover the essential requirements of the standard. This is not an easy task and it is often very much in the hands of the working group chairman to find ways to get a common understanding. In some cases it also takes some time to find the right wording to obtain consensus in the working group. As long as political issues or particular company policies can be avoided, experience suggests that a consensus text can be found.

Clause 7.3 of DIN 820 defines the process of the public appeal for comments. The draft text of the standard must, before final publication, be published to ask for public comments. All collected comments must be discussed and evaluated in the appropriate working group of the technical committee. For this reason, all commenters must be invited to present their position. Each commenter can start an appeal process to find agreement on the comment.

This public appeal process is not used so often; in most cases the commenters accept the decision of the technical committee evaluating the comment and the result on what to do or not to do with the standard text. In some cases electronic meetings are used to clear the case.

Clause 7.3 of DIN 820 also outlines how and when the standard needs to be sent for public comments and how much time (in general 2–5 months) can be given to send comments to the technical committee. After the final deadline has passed, no more comments will be accepted on the standard.

5.2.10   Benefits of Standards

From a German point of view, standards are seen as tools to open and create international market places for the country's export industry with a safety level that is as high as possible and a quality level that is as high as is economically possible. Box 5.2 gives the main advantages of standards. Standards are a way to reduce worldwide barriers to market access for products and services. An active standard revision process usually encourages innovation towards higher technical performance at lower product cost for the user. In this way, standards offer benefits for society in an innovative surrounding.

The great advantage of standardization is compatibility of products and systems. Technical solutions from different manufacturers are interchangeable. Take a household light switch, for example. There are many manufacturers of such products and a large variety is available on the market. They all fit into the same holder system. So users have a free choice of the brand they like. From a German industry point of view, this encourages innovation and competitive industries with high level of safety and quality products.

A big benefit comes from the product test requirements. Standards give precise technical requirements for type tests, routine tests, factory tests and on-site tests to guarantee the highest safety and quality level. It would be very inefficient and costly if each manufacturer were to come up with its own test requirements and whether these tests would fulfil the need for safe and reliable products and systems would be questionable.

Box 5.2   Benefits of standards – a German view.

Remove trade barriers
Support technical innovation
Provide system compatibility
Set basics for product testing
Increase safety of products and systems
Increase speed to market for innovations
Provide a knowledge platform for industry, science, the public and the state

→ **Estimated annual profit from standards: 16 billion euros per year**

Standards provide safe products and systems, which benefit society. This is reflected in the relatively low number of cases of injury or damage caused by products. Improvements resulting from standards lead to lower numbers of accidents as statistics over recent decades show, even with ever increasing numbers of electrical products and systems.

Germany, as a technology-driven nation, sees standards as beneficial in that they increase the speed to market for innovative products. Germany is one of the nations most strongly engaged international standardization. Experts from industry are contributing to this with their innovative knowledge and standardization work, sponsored by industry. They are consulted on technical questions, are part of the discussions and they have contact with the government.

5.2.11　Adoption of CENELEC Documents

The CENELEC documents use the same numbers in the German DIN/V DE standards as in CENELEC, as shown in Table 5.6.

The CENELEC documents in national versions of the member nations do have the same structural concept. They start with a national title page, followed by a national preface and, if necessary, a national annex before the official part follows.

The national title page gives information about titles and numbers and how they are related to the CENELEC document. The national preface gives information about possible changes with the withdrawal of national standards. If there is a national deviation, the differences from the standard text are given in the national annex to the document. In this national annex, explanations or deviations from the document's text are explained and information on national adaptations is given.

In Germany all CENELEC standards are translated into the German language. For EN standards the rule is that any document will be translated into German while, for EN technical specifications and technical reports, the national committee of Germany decides on a case-by-case basis. The valid version in case of doubt is always the English version of the EN standard.

Table 5.6　Number and title of CENELEC documents in German DIN/VDE.

European	German DIN/VDE
EN 50000:2015 Title xxxx	DIN EN 50000 Title xxxx German version EN 60000:2015 Or if safety related: DIN VDE EN 50000 Title xxxx
EN 60000:2015 Title xxxx (IEC 60000:2014)	DIN EN 60000 Title xxxx (IEC 60000:2014) German version EN 60000:2015 Or if safety related: DIN VDE EN 50000 Title xxxx (IEC 60000:2014) German version EN 60000:2015
HD 355 S1:2014 Title 2014	DIN 4577 Title xxxx German version HD 355 S1:2014 Or if safety related: DIN VDE 0111 (VDE0111) Title xxxx

EN draft standards are marked by the abbreviation 'pr' (French *project*) – so this appears as 'prEN'. The prEN standards are not translated into German. The numbering is as shown in Table 5.6.

The numbering system for CENELEC documents used in Germany identifies the document as having originated in CENELEC when the number of the document is in the 50000 series. In case of documents that originated in IEC or ISO the 60000 series is used. If the title of the document refers to IEC and ISO then the EN document is identical to the IEC or ISO document. If no reference in the numbering system is made to a specific IEC or ISO document then the EN document has been modified. In most cases the IEC or ISO and EN documents are identical.

Safety-related standards are seen as all electrical standards and they will receive a classification with VDE in the document number. All other standards are identified as DIN (www.normenbibliothek.de/index.html, accessed 7 February 2016) VDE and DIN are publishing handbooks on how to use standards. In VDE they are called 'VDE-Anwendungsregeln' and in DIN they are called DIN SPEC. They are published by Beuth Verlag (http://www.beuth. de/en, accessed 2 March 2016).

5.2.12   Figures and Numbers

In Germany there are four sources of standards today. The first source is the DIN – these are national documents. The second is direct adaptation from the IEC. The third is a pure European document and the fourth originated in the IEC and was taken by CENELEC as an EN and then into a national German document.

The pure national standards of DIN make up less than 5% of all standards valid in Germany. In 1990 the share was 40% and it was even higher before that. The harmonization of national standards into EN standards made this reduction possible. The same can be said about the direct adaptation of IEC standards into German DIN documents – down from 40% in 1990 to less than 5% today.

Pure European standardization work is still in the range of 10% and has been relatively constant over the years. There is a partial need for European standards where there is no interest at the international level of the IEC. The great majority of German standards are based on IEC documents transferred to EN documents and then into German DIN/VDE documents. This share increased from zero before CENELEC in the 1980s to about 90% of all valid standards in Germany today.

The monthly periodical, *DIN-Mitteilungen,* informs the public about any standardization activity in DIN/DKE, CENELEC and ISO that also can be found at the DIN web site (www.din-mitteilungen.de, accessed 21 February 2016). The DIN-VDE Standards Library (www. normenbibliothek.de, accessed 21 February 2016) is the Internet platform where comments can be made on DIN-VDE, CENELEC or IEC standards. The standards can be seen for free at several offices in Germany – see the Beuth web site, www.beuth.de (accessed 21 February 2016).

5.2.13   VDE Application Guide

The VDE application guide (*VDE-Anwendungsregeln*) process (www.vde-verlag.de/normen/ anwendungsregeln.html, accessed 7 February 2016) is defined in VDE 0022. The process is designed for fast-changing, dynamic, technical markets – for example electromobility.

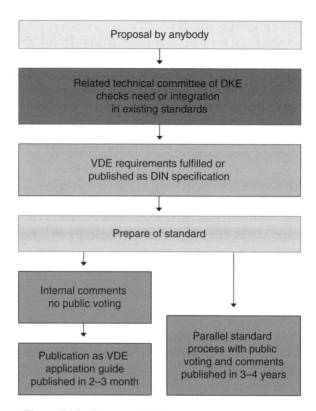

Figure 5.15   Process to initiate VDE Application Guide.

It combines standardization with a quick reaction to fast-changing technologies by reducing the publication time of the regular standardization process. Anyone may initiate the publication of a *VDE-Anwendungsregeln* within 3 months. Before publishing, the criteria of the VDE 0022 process need to be fulfilled as shown in Figure 5.15.

The process separates the technical work on the new standard into two parallel tracks. The fast track will only check the need for such a standard or if the content can be integrated into an existing standard. The *VDE-Anwendungsregeln* can be published after 2–3 months' internal draft work with comments from the related technical committee of DKE (www. vde-verlag.de/normen/anwendungsregeln.html, accessed 7 February 2016).

In parallel to this work, the normal public standardization process will be carried on within the related technical committee. The standard will then be developed and published typically within 3 to 4 years. Once the standard is published the *VDE-Anwendungsregeln* will be withdrawn.

Today there are more than 40 *VDE-Anwendungsregeln*s in the technical fields of photovoltaic technology, metering, building technology, ambient living, energy efficiency and energy networks.

5.2.14   VDE Specification Code

In the VDE there are five different types of documents available as shown in Table 5.7.

Table 5.7   Types of VDE documents.

VDE 0xxx	VDE standards (*VDE Bestimmungen*)
VDE V 0xxx	VDE prestandards (*VDE Vornormen*)
VDE 0xxx	VDE rules (*VDE Leitlinien*)
VDE AR N 4xxx	VDE application guides (*VDE Anwendungsregeln*)
VDE 0xxx Bbl	VDE annexes (*VDE Beiblätter*)

The VDE standards (*VDE-Bestimmungen*) are the core elements of the total standard documents. They are supported by VDE prestandards (*VDE-Vornormen*), which have interim validity until the final standard is available. This reflects a changing and innovative technical market. The VDE rules (*VDE-Leitlinien*) give information about how to write a good standard. The VDE application guide (*VDE-Anwendungsregeln*) is a fast way to publish a specification and work in parallel on the full standardization process with public comments and voting. The VDE annexes (*VDE-Beiblätter*) cover additional information for the safe and correct use of standards or adapt the content of standards to technical innovations.

DIN SPEC

The DIN specification (DIN SPEC) is a process to publish documents that do not fulfil all the requirements of a standard – for example, they might contain conflicts with other standards.

A DIN SPEC can come from a VDE prestandard, from an internal report of a technical committee, from an IEC publicly available specification (IEC PAS) or a CEN or CENELEC workshop result (CWA).

5.2.15   The DKE's View of Standardization in the United States

The DKE sees ANSI as a 'holding' organization for American national standards and the national member in IEC and ISO. Unlike Germany, which has a unique standardization system with the DIN and DKE, in the United States the standardization has a decentralized structure with about 600 organizations active in standardization, with different technical focuses and representing different technical or administrative organizations.

An additional 270 standards-developing organizations (SDO) are accredited by ANSI. Such standard-writing organizations include, for example, ASTM for testing and materials, NEMA for electrical manufacturers, ASME for mechanical engineering, and the IEEE for electrical and electronic engineering. Any of these organizations has international activities in competition to IEC and ISO.

There is no coordination of standardization activities in the United States comparable to the DIN or DKE in Germany. For this reason different specifications on the same technical area are in the market and unified testing or certification is not possible.

There is also no systematic transfer from IEC or ISO standards as in EN with the Dresden and Vienna Agreement. The dual logo agreement between IEC and IEEE is one of few coordination of standardization.

The DKE is not directly connected to the 270 SDOs but keeps contact, case by case, when topics arise that are of importance in the United States and internationally. One contact, for example, is established on an informal basis between the DKE and IEEE.

5.3   France – AFNOR

5.3.1   History

The Association Française de Normalisation (AFNOR) is an international services-delivery network with four core competence areas: standardization, certification, industry press releases and training (www.afnor.org, accessed 8 February 2016).

Founded in 1926, AFNOR is a public nonprofit organization serving 60 000 customers in more than 100 countries. With a work force of 1373 employees, including 350 employees abroad (figure from 2015), AFNOR operates 13 regional delegations and 39 agencies worldwide. There are more than 20 000 experts enrolled in active standardization work with AFNOR, mostly active in the IEC, ISO, CEN and CENELEC on the international and European level. There are 1854 translators (2015 figures) working on French translations of international standards.

The values of AFNOR are linked to providing high-quality standards for industry, based on wide trust as a third-party independent standardization organization; to being responsible for efficient standardization work processes; and to providing a platform for open and transparent committees in standardization.

5.3.2   Structure

AFNOR is active in French standardization (NF), European standardization (EN) and international standardization (ISO and IEC standards). It is the French member of European and international nongovernmental standardization organizations and acts as a national committee. Based on the strong industrial position of France AFNOR plays an important role in international standardization.

Supported by French industry, experts are active participating in the strategy, organization and technical work in national and international standardization.

The French governmental cross ministry of economy, finance and sustainable development (ADETEF) supports AFNOR in international standardization projects. The objective of this international engagement is to maintain the quality of products and services at a high level and to promote economic development and increasing competitiveness on the world market. AFNOR is cooperating and following the World Trade Organization (WTO) in the goal of removing international trade barriers between countries. It offers the experts a high-performing infrastructure for the standardization process. With this, AFNOR performs as a network for experts and offers multiple activities.

AFNOR has a strong link to European standards with more than 90% of the French standards originating as European standards (EN), most without any deviation. For this AFNOR delegates many experts to European standardization work and plays an active role in CEN and CENELEC.

On an international level, France played a leading role in ISO and IEC as a member of the Technical Management Board in 2015 and is the third largest contributing member out of 163 countries. In standardization and certification, AFNOR has a privileged relationship with China, Brazil, Algeria and Macedonia.

AFNOR was founded, together with the Bureau de Normalisation de Québec (BNQ) and the Réseau Normalisation et Francophonie (RNF), to support the French language and standards worldwide, with activities in Africa, North America and Europe.

Table 5.8   Standardization process in France.

1	Request	Any French socio-economic player
2	Assessment	Sector-based standardization bureau of AFNOR decides to start or not
3	Program of work	Secretary of the standardization commission informs CEN/CENELEC
4	Draft	Draft as proposal sent to the standardization commission
5	Consultation	Secretary of the standardization commission reaches consensus
6	CEN/CENELEC inquiry	– Proposed draft will be a standard
		– Results of comments evaluated
7	Certification and publication	Standards are published by AFNOR:
		– NF ISO International standard or applied in France NF IEC
		– NF EN ISO French standard of international origin applied in Europe and France NF EN IEC
		– NF EN French standard of European origin
		– NF Partly French standard

AFNOR is a sector-based organization, covering a wide range of technical fields. It is administered by a board of directors with 30 representatives from French industry, trade, consumers and NGOs. The standardization work is coordinated by the Standardization Coordination and Steering Committee (CCPN).

The Audit and Evaluation Committee (CAE) audits and evaluates the French standardization system. The sector-based structure of AFNOR provides a single expert for each technical field. These experts in AFNOR act as true project managers of related standardization projects – national, European or international – and make sure that French interests are taken into account.

The standardization process in France is shown in Table 5.8. The request for new standardization work can come from any social or economic player in France. Within AFNOR the related technical sector expert group will decide if a project on standardization will be started or not. If it is approved, the work is added to the French program of work on standardization by the secretary of the standardization commission and information of this new French work will be given to CEN or CENELEC. A draft proposed by the working group will be sent to the standardization commission and the secretary will obtain consensus on the draft.

An inquiry on the proposed standardization work will be made to CEN or CENELEC to follow the European Directives if the standard is from interest for Europe or will stay in France.

The comments received on the draft will be evaluated and addressed. AFNOR is the final publishing organization in France, whether the standard will be an international one from ISO or IEC, a French standard of international origin, a French standard of European origin, or a purely French standard.

The organization of AFNOR into 13 sectors is shown in Table 5.9. In total more than 20 000 experts are participating in French international and national standardization work in more than 1300 technical committees.

Table 5.9   Active sectors.

Food and Farming	81 commissions
Aeronautics and Aerospace	24 commissions
Automotive	32 commissions
Banking/Insurance/Finances	8 commissions
Construction Trades	206 commissions
Water and Sanitation	86 commissions
Energy	53 commissions
Digital	82 commissions
Industry	466 commissions
Health Medical Social Care	66 commissions
Services	99 commissions
Sport and Leisure	37 commissions
Transport and Logistic	111 commissions

5.4   Spain – AENOR*

5.4.1   History

AENOR (www.aenor.es, accessed 8 February 2016) was founded in 1986, coinciding with Spain's accession to the European Economic Community. The opening of borders that this represented was at once a huge opportunity and a tremendous challenge for Spanish products.

Prior to that date, standardization was the responsibility of the Spanish Institute of Rationalization and Normalization (IRANOR), a public company created in 1945 under the Centre for Higher Scientific Research. Spain's economic and public administration sectors agreed the country needed a similar organization to those already existing in other European countries. Thus, all standardization and certification activities moved to the private sector with the setting up of a private, nonprofit, independent organization. The objective was to spread the culture of quality throughout the Spanish productive fabric in order to improve its competitiveness.

First Steps in Standardization

The first year saw the creation of the first 24 technical standardisation committees, mostly as a result of the transfer of IRANOR's technical operations, and a basic body of 7810 standards was established, also inherited from this organization. One year later, AENOR began representing Spain before European bodies (CEN, CENELEC and ETSI) and international organizations (ISO and IEC).

There are currently more than 200 technical standardization committees, in which nearly 6000 experts participate. Their work is internationally recognized, as demonstrated by the fact that standards developed in Spain are increasingly used as a benchmark for the creation of European and international standards.

---

*Enrique Otegui contributed to this section.

Certification Work

In 1986, the basic infrastructure was put in place to offer certification services, which at that point were confined exclusively to product certification. The first technical certification committees were for plastics and household appliances. More and more committees were gradually formed, until the current figure of 70 was reached. In the first decade, product certification was used for construction and electrical equipment. It was only at the end of the 1990s that it was extended to new fields such as food, handicrafts and services.

In 1989, AENOR began to certify management systems in accordance with the UNE-EN ISO 9001 standard and this activity has continued to develop, surpassing the 26 000 certificate mark. In the 1990s, this certification was predominantly required by purely industrial organizations but, following the turn of the century, with the publication of a new version of the ISO 9001 standard in the year 2000, the remit was extended to include service organizations and SMEs.

The 1992 Earth Summit, held in Rio de Janeiro, heralded the beginning of the development of environmental policies. AENOR began to work in this field that same year, before issuing its first environmental management system certificates in 1994. The second most implemented management system, AENOR has issued 6900 of these certificates.

25th Anniversary

In 2011, 25 years after it was founded, AENOR had 200 technical standardization committees. The Spanish technical standards catalogue includes more than 28 000 standards. More than 90 000 products feature the N mark and more than 60 000 management system certificates have been issued. AENOR is among the ten most important certification organizations in the world.

AENOR is one of the success stories of the Spanish economy. Moreover, this is a shared achievement, as its contributions to the wellbeing of all have been achieved thanks to the participation of tens of thousands of contributors to the economy, such as the AENOR member organizations, all levels of public administration, experts from the standardization and certification committees and certified organizations.

Landmarks

Important landmarks in the history of AENOR are outlined in Table 5.10.

5.4.2   AENOR's Profile

The activity of AENOR helps improve the quality and competitiveness of companies, their products and services through the development of technical standards and certification, thereby helping organizations to generate one of the most sought-after values in today's economy: trust.

It is the organization that is legally responsible for developing and disseminating technical standards in Spain. Standards indicate what a product should be like or how a service should work so that it is safe and meets the consumer's expectations. AENOR offers one of the most extensive catalogues available, with more than 28 900 standard-related documents containing effective solutions.

Table 5.10   Landmarks in the history of AENOR.

1986	AENOR is founded
1987	First UNE standard edited
1988	First N mark product certificate
1989	First Quality Management System certificate
1990	First International Cooperation project
1992	Training centre created
1993	The first delegation opens, situated in the Basque Country, and the first AENOR Agency opens in Andalusia
1995	First Environmental Management System certificate
1996	AENOR, first Spanish company accredited by the Spanish National Accreditation Body (ENAC)
1997	AENOR México opens
1999	CEIS (Testing, Innovation and Services Centre) created
2001	AENOR internacional created and AENOR Chile opens
2002	First RDI Management System certificate
2002	AENOR Italia opens
2004	First Management System certificate in Occupational Health and Safety
2004	AENOR Centroamérica opens
2005	Accreditation by United Nations as a Designated Operational Entity – Kyoto Protocol. First Spanish company to achieve it.
2005	LUSAENOR (Portugal) and a Bulgaria office open
2006	AENOR Brasil opens
2007	AENOR Perú opens
2008	AENOR laboratorio created
2009	Opening of the latest delegation, Castile-La Mancha. AENOR present in all 17 autonomous communities
2009	AENOR Polska and offices in the Dominican Republic and Morocco open
2010	AENOR Ecuador opens
2010	First goods inspection
2011	First certificate of conformity recognition agreement with third party countries (Saudi Standards, Metrology and Quality Organization – SASO)
2012	Consultancy Service on Technical Standards and Legislation (SAT) and AENORmás launched
2013	Spain exceeds 30 000 standards in catalogue
2013	EFR certificate acquired
2014	Adherence to the Global Compact

Certification

The serious and rigorous work that has defined the company since it was founded has helped make AENOR's certificates the most highly valued – not just in Spain but around the world, with the organization having issued certificates in more than 60 countries.

To spread the culture of quality, AENOR also engages in significant publishing activity, it designs software for system management and it delivers special training and offers various information services.

In the field of testing, it is the senior partner of CEIS (Testing, Innovation and Services Centre), an international benchmark centre whose activity includes the performance of compliance tests, technical studies and preventive and predictive maintenance of facilities.

Likewise, 2008 saw the creation of AENOR laboratorio, aimed at all the players in the food sector, such as primary producers, processing, distribution and services industries. It covers the three principal areas of analysis: physicochemical, microbiological and sensory.

AENOR offers a proven track record and information on standards and products and services connected with organizations from all over the world, performing important work in the field of international cooperation.

In Spain, AENOR is present in all the autonomous communities, where it has 20 offices, and has a permanent presence in 12 countries, mostly in Latin America and Europe.

5.4.3 AENOR Structure

In this section an overview is given of how AENOR is linked to the IEC and CENELEC technical committees. AENOR covers the following technical fields in international standardization.

Basic Electrical Standards

Standardization of:

- information structures, documentation and graphical symbols;
- basic principles and safety principles for man-machine interface, marking and identification;
- quantities and units and their letter symbols;
- dependability;
- magnetic alloys and steels;
- electrostatics and safety of electrostatic painting and finishing equipment;
- environmental conditions, classification and methods of testing;
- general environmental considerations for electrical and electronic products and systems;
- functional safety;
- electrotechnical guides;
- degrees of protection provides by enclosures;

excluding:

- environmental considerations for products within the scope of other technical committees.

See Table 5.11 for international relationships in this field.

Low-Voltage Switchgear, Controlgear and Accessories

Standardization of:

- low-voltage switchgear and controlgear;
- low-voltage switchgear and controlgear equipment;
- electrical accessories (plugs, socket-outlets and switches, cable management systems, connection devices);
- fire hazard testing;

Table 5.11   International relationships – basic electrical standards.

AENOR structure	International relationships	
	IEC	*CLC*
SC 3	TC 3	TC 111/SC X
SC 16	TC 25	TC 204
SC 25	TC 56	
SC 56	TC 68	
SC 68	TC 70	
SC 70	TC 101	
SC Guides	TC 104	
SC SF	TC 111	
SC 101-204		
SC 104		
SC 111		

Table 5.12   International relationships – low-voltage switchgear, controlgear and accessories.

AENOR structure	International relationships	
	IEC	*CLC*
SC 17BD	TC 23	BTTF 52-3
SC 23	T 89	BTWG 112-1
SC 23S-213	TC96	BTTF 146-1
SC 23BG	TC 121	TC 23/SC BX
SC 23C		TC 23/SC E
SC 23E		SR 121
SC 23F		TC 121/SC A
SC 23H		TC 123
SC 23J		
SC 89		
SC 96		

- low-voltage and small power transformers, reactors, power supply units and similar products;
- equipment for cable trunking, ducting and support.

See Table 5.12 for international relationships in this field.

Electrical Installations

Standardization of:

- electrical apparatus for explosive atmospheres;
- electrical installations and protection against electric shock;

Table 5.13 International relationships –
electrical installations.

AENOR structure	International relationships	
	IEC	*CLC*
SC 31	TC 31	JWGCMI
SC 64	TC 64	TC 31
SC 205		TC 64
SC 216		TC 205
SC 218		TC 216
GT 1		TC 218

- home and building electronic systems;
- gas detectors.

See Table 5.13 for international relationships in this field.

Electric Equipment and Automatic Systems for Industry

Standardization of:

- rotating electrical machines;
- electric equipment for railways;
- secondary cells and batteries;
- power electronics;
- electric welding;
- industrial electro heating;
- power capacitors;
- primary cells and batteries;
- safety of machinery: electrotechnical aspects;
- safety of hand-held and transportable motor-operated electric tools;
- industrial-process measurement and control;
- electric road vehicles and electric industrial trucks;
- alarm systems;
- surface assembly technology and printed electronics;
- equipment for electrotechnical surface transport systems: electrotechnical aspects.

See Table 5.14 for international relationships in this field.

Electrical Safety

Standardization of:

- equipment and tools for live working;
- safety of people in high-voltage installations.

See Table 5.15 for international relationships in this field.

Table 5.14 International relationships – electric equipment and automatic systems for industry.

AENOR structure	International relationships		
	IEC	ETSI	CLC
SC 2	TC 2	RT	BTTF 60-1
SC 9X	TC 9		BTTF 132-2
SC 21	TC 21		TC 2
SC 22	TC 22		TC 9/SC XA
SC 26	TC 26		TC 9/SC XB
SC 27	TC 27		TC 9/SC XC
SC 33	TC 33		TC 9/SC X
SC 35	TC 35		TC 21/SC X
SC 44	TC 44		TC 22/SC X
SC 65	TC 65		TC 26/SC A
SC 69	TC 69		TC 26/SC B
SC 79	TC 79		TC 44/SC X
SC 91	TC 91		TC 65/SC CX
SC 116	TC116		TC 69/SC X
	TC 119		TC 79
			TC 116

Table 5.15 International relationships – electrical safety.

AENOR structure	International relationships	
	IEC	*CLC*
SC 78	TC 78	BTTF 62-3
SC BTTF 62-3		TC 78

Table 5.16 International relationships – lamps and related equipment.

AENOR structure	International relationships		
	IEC	*CLC*	*CEN/CLC*
SC 34	TC 34	BTTF 142-1	CG-LIGHT
SC 34A		TC 34/SC A	
SC 34B		TC 34/SC Z	
SC 34C			
SC 34D			

Lamps and Related Equipment

Standardization of lamps and related equipment: lamp caps, lamp holders, gauges, ballasts, luminaries and auxiliaries for lamps, emergency lighting systems, road traffic lighting and signalling.

See Table 5.16 for international relationships in this field.

Electrical Energy Production

Standardization of:

- hydraulic turbines;
- steam turbines;
- electrical insulating materials;
- nuclear instrumentation;
- solar photovoltaic energy systems;
- wind turbines;
- superconductivity;
- fuel-cell technologies;
- power-generation equipment procurement guides;
- wave, tidal and other current converters;
- solar thermoelectric energy systems.

See Table 5.17 for international relationships in this field.

Transmission and Distribution of Electrical Energy

Standardization of:

- overhead electrical conductors;
- overhead lines;
- equipment for electrical energy measurement and load control;
- power transformers;
- high-voltage switchgear and controlgear;
- high-voltage enclosed switchgear and controlgear;
- insulation co-ordination;
- fuses, electrical insulators;
- surge arresters;

Table 5.17   International relationships – electrical energy production.

AENOR structure	International relationships		
	IEC	*CLC*	*CEN/CLC*
SC GC	TC 4	TC 45/SC A	JWGFCGA
SC 004	TC 5	TC 82	TC 2
SC 005	TC 15	TC 88	
SC 015	TC 45		
SC 045	TC 82		
SC 082	TC 88		
SC 088	TC 90		
SC 090	TC 105		
SC 105	TC 114		
SC 114	TC 117		
SC 117			

- instrument transformers;
- high-voltage testing techniques;
- power systems management and associated information exchange;
- short-circuit currents;
- lightning protection;
- all-or-nothing electrical relays;
- measuring relays and protection equipment;
- electrical insulation systems, system engineering and erection of electrical power installations in systems with nominal voltages above 1 kV AC and 1.5 kV DC;
- mains communicating systems;
- fluids for electrotechnical applications;
- high-voltage direct current (HVDC) transmission for DC voltages above 100 kV.

See Table 5.18 for international relationships in this field.

Table 5.18 International relationships – transmission and distribution of electrical energy.

AENOR structure	International relationships		
	IEC	*CLC*	*CEN/CLC*
SC 7	PC 118	BTTF 128-2	SMCG
SC 10	TC 7	BTTF 129-1	
SC 11	TC 10	BTTF 132-1	
SC 13	TC 11	TC 11	
SC 14	TC 13	TC 14	
SC 17C	TC 14	TC 17/SC AC	
SC 17-17A	TC 17	TC 36/SC A	
SC 28	TC 28	TC 37/SC A	
SC 32B	TC 32	TC 38/SC X	
SC 32C	TC 36	TC 57	
SC 32-32A	TC 37	TC 81/SC X	
SC 36	TC 38	TC 94	
SC 37	TC 42	TC 99/SC X	
SC 38	TC 57	TC 205/SC A	
SC 42	TC 73		
SC 57	TC 81		
SC 73	TC 94		
SC 81	TC 95		
SC 94	TC 99		
SC 95	TC 109		
SC 99	TC 112		
SC 109	TC 115		
SC 112			
SC 115			
SC 205A			

Electromagnetic Compatibility

Standardization of electromagnetic compatibility (EMC), EMC products, avoidance of radio disturbances. See Table 5.19 for international relationships in this field.

Electronic Equipment

Standardization of electroacoustics, electronic tubes, capacitors and resistors for electronic equipment, semiconductor devices, electromechanical components and mechanical structures for electronic equipment, piezoelectric and dielectric devices for frequency control and selection, magnetic components and ferrite materials, printed circuits, electrical equipment in medical practices, safety and energy performance of information technology equipment, laser equipment, ultrasonics, safety of audio, video and analogue equipment, entertainment and educational electronic equipment for domestic and analogue use. See Table 5.20 for international relationships in this field.

Table 5.19 International relationships – electromagnetic compatibility.

AENOR structure	International relationships	
	IEC	CLC
SC CISPR-210 SC 77-210	CISPR TC 77	TC 210

Table 5.20 International relationships – electronic equipment.

AENOR structure	International relationships			
	ISO	IEC	CLC	CEN/CLC
SC 29	TC 172/C 9	TC 29	BTTF 116-2	TC 123
SC 39		TC 40	JWGAIM	
SC 40		TC 47	TC 40/SC XA	
SC 47		TC 48	TC 40/SC XB	
SC 48		TCC 49	TC 62	
SC 49		TC 51	TC 76	
SC 51		TC 62	TC 107/SC X	
SC 52		TC 76	TC 108/SC X	
SC 62		TC 87		
SC 76		TC 107		
SC 87		TC 108		
SC 93-217		TC 110		
SC 107				
SC 108				
SC 116-2				

Electrotechnical Aspects of Telecommunication Equipment

Standardization of maritime navigation and radio communication equipment and systems, audio, video and multimedia systems and equipment, electrotechnical aspects of telecommunication equipment, equipment used in mobile services and satellite communications systems, broadcasting transmitters, broadcasting receivers, data broadcasting systems, cable networks for television signals, sound signals. See Table 5.21 for international relationships in this field.

Electric Energy Cables

Standardization of:

- constructive requirements, testing and end-use recommendations for conductors and insulated electrical power cables and their accessories, for use in wiring and in power generation, distribution and transmission;
- electric cables specifically designed for marine applications and their installations;
- wound wires;

excluding

- cables for communication and
- data-transmission cables.

See Table 5.22 for international relationships in this field.

Table 5.21   International relationships – electrotechnical aspects of telecommunication equipment.

AENOR structure	International relationships		
	ISO	*IEC*	*CLC*
SC 80	JTC1/SC 5	TC 80	BTTF 133-1
SC 100		TC 100	TC 100/SC X
SC 206		TC 103	TC 209
SC 209			TC 215
SC 215			

Table 5.22   International relationships – electric energy cables.

AENOR structure	International relationships	
	IEC	*CLC*
SC 0A	TC 18	TC 18/SC X
SC 18A	TC 20	TC 20
SC 20A	TC 55	TC 55
SC 20B		
SC 55		

Table 5.23　International relationships –
telecommunication cables and optical fibre.

AENOR structure	International relationships	
	IEC	*CLC*
SC 46	TC 46	TC 46/SC XA
SC 86	TC 86	TC 46/SC XC
		TC 46/SC X
		TC 86/SC BA
		TC 86/SC A

Table 5.24　International relationships –
household electrical appliances.

International relationships		
IEC	*CLC*	*CEN*
TC 59	TC 59/SC X	TC 429
TC 61	TC 61	
TC 72	TC 72	

Telecommunication Cables and Optical Fibre

Standardization of cables, wires, waveguides, connectors and accessories for communication and signalling, optical fibre. See Table 5.23 for international relationships in this field.

Household Electrical Appliances

Standardization of:

- performance of household electrical appliances;
- safety of household and similar electrical appliances;
- automatic controls for household use.

See Table 5.24 for international relationships in this field.

Electrical Installations for Lighting and Beaconing of Aerodromes

Standardization of electrical installations for lighting and beaconing of aerodromes. See Table 5.25 for international relationships in this field.

Equipment and Methods for the Assessment of Electromagnetic Fields Associated with Human Exposure

Standardization of measurement equipment and methods of electromagnetic fields from $0\,Hz$ to $300\,GHz$ associated with human exposure. See Table 5.26 for international relationships in this field.

Table 5.25   International relationships – electrical installations for lighting and beaconing of aerodromes.

International relationships	
IEC	*CLC*
TC 97	TC 97

Table 5.26   International relationships – equipment and methods for the assessment of electromagnetic fields associated with human exposure.

International relationships	
IEC	*CLC*
TC 106	TC 106/SC X

Renewable Energies, Climate Change and Energy Efficiency

Standardization in the field of organization and definition of:

- promotion tools for renewable energy production;
- promotion tools for greenhouse gas emissions trading;
- tools for the profitable improvement of energy end-use efficiency

including

- definitions and terminology of renewable energies, climate change and promotion of energy end-use efficiency and energy services;
- guidelines, recommendations, requirements and specifications;
- auditing guides for certification;
- energy management systems;
- energy efficiency and saving calculation

but excluding

- all those elements and systems standardized in AEN/CTN 206 Electric Energy Production;
- all those standardization works related to energy performance in building developed by other standardization committees.

See Table 5.27 for international relationships in this field.

Table 5.27 International relationships – renewable energies, climate change and energy efficiency.

AENOR structure	International relationships		
	ISO	*CEN/CENELEC*	*CEN*
GT 2	TC 203	SFEM	BT/WG 210
GT 3	TC 207/SC 7	BT/TF 190	
GT 4	TC 242	JWG 1	
GT 5	TC 257	JWG 2	
GT 6	TC 265	JWG 3	
GT 7		JWG 4	

Table 5.28 International relationships – electrical energy supply systems.

International relationships	
IEC	*CLC*
TC 8	BTWG 143-2
	TC 8/SC X

Electrical Energy Supply Systems

Standardization of electrical energy supply systems operation and the preparation, in cooperation with other committees, of the development of standards needed to facilitate an electrical energy supply of quality for consumers in the open market taking into account the following aspects:

- terminology;
- quality of supply;
- connection to the grid;
- electrical sector related services

excluding

- equipment, products or installations that are within the competence of other committees.

See Table 5.28 for international relationships in this field.

Electrical Energy Storage (EES) Systems

Standardization in the field of grid-integrated EES systems focusing on system aspects on EES Systems rather than energy storage devices. See Table 5.29 for international relationships in this field.

Table 5.29   International relationships –
electrical energy storage systems.

International relationships – IEC
TC 120

5.5   Italy – CEI/UNI

5.5.1   History

In Italy there are two standardization organizations: the Italian Electrotechnical Committee (CEI) for electrical equipment and services and Ente Italiano di Normazione for anything else (UNI). See www.ceiweb.it (accessed 22 February 2016) www.uni.com (accessed 22 February 2016) and https://www.ihs.com/products/uni-standards.html (accessed 8 February 2016).

Italian Electrotechnical Committee (CEI)

The Italian Electrotechnical Committee is a private, nonprofit association, responsible at the national level for standardization in the electrotechnical, electronic and telecommunication fields. It is mandated by the Italian state for activities in Europe with CENELEC and internationally with the IEC.

Founded in 1906 and officially recognized by the Italian government and by the European Union, CEI publishes and disseminates technical standards based on Italian Law 186/1968.

The Italian Electrotechnical Committee is the national standardization organization and represents Italy in CENELEC for European Standardization following the European Directives and Regulations and in the IEC for international standardization.

The CEI not only publishes technical standards in Italy – it also promotes research and the development of technology and knowledge for the standardization process and it is involved with social issues such as health, safety and environmental protection. The following are the CEI's main fields of interest:

- publication of national, European and international standards;
- definition of requirements, setting test procedures and providing evaluation criteria for test laboratories;
- co-ordinating stakeholder focus points;
- participation in technical research and development;
- promotion of conformity assessment;
- promotion of a standardization culture and offering training;
- participation in European and international standardization with a mandate from Italy.

Statutory Bodies

The Italian Electrotechnical Committee (CEI) has a general assembly, a council and an executive committee with the functions of president general, managing director, auditor and technical committees.

General Assembly

The general assembly is formed by the members of CEI who have an interest in technical standardization as full members or affiliate members. There are three sponsoring members:

- the Italian Association for Electrotechnical, Electronics, Automation, Informatics and Telecommunication (AEIT);
- the National Electrical and Electronic Manufacturers Federation (ANIE);
- the National Electric Energy Supplier (ENEL).

The Ministry of Economic Development sponsors and sets measures of control at policy level through its position in the executive committee.

Council

The council has as representatives of the sponsoring members on behalf of the general assembly, two representatives from the Ministry of Economic Development, 11 members from other ministries (one each), four members from the National Research Council and 12 elected members from the General Assembly.

Executive Committee

The executive committee consists of the president, four vice-presidents, two members elected by the council and the secretary. It reports to the council, which approves its finances and technical program.

Technical Bodies

The technical committees and subcommittees are in charge of the development of the normative documents for each sector. The activities are carried out by the experts nominated by the CEI.

5.6 United Kingdom – BSI

5.6.1 General

The British Standards Institution (www.bsigroup.com, accessed 7 February 2016) is the national body for national and international standardization in the United Kingdom. Founded in 1901 by Sir John Wolfe Barry, the designer of London's Tower Bridge, the standardization activities started with reducing the variety of sizes of structural steel sections. The British Standards Mark known as the Kitemark was introduced in 1903 as a marker for buyers.

During the First World War the British Standard was used by the Admiralty and many institutions in the United Kingdom. In 1920 the British Standards Institute went international and spread to Canada, Australia, South Africa and New Zealand and developed interest in the United States and Germany. Its current name, 'The British Standards Institution' was adopted in 1931. During the Second World War, normal standardization activities were stopped and more than 400 emergency standards were published until 1945.

In 1946 the BSI organized the Commonwealth Standards Conference, which was the basis for the foundation of the International Organization for Standardization (ISO). In the 1950s and 1960s the standardization market was flooded with consumer goods and the Kitemark was applied to furniture, cookers, motorcycle helmets and many other items. In 1959, the Test House at Hemel Hempstead started to test equipment produced for Canada.

Between 1975 and 1997 the United Kingdom installed a quality assurance system. The principles were defined in BS 5750 from 1979, which was superseded later by ISO 4000 in 1987. In 1992 the BSI published the world's first environmental management system with BS 7750, which was superseded by ISO 14001 in 1996. Both standardization activities have been very successful around the world with more than 1 million ISO 9001 certificates and more than 220 000 ISO 14001 certificates in 2009.

The international activities of BSI were been strengthened in 1991 with an office in Herndon, Virginia, United States and in 1995 in Hong Kong.

Since 1998, the BSI has gone international with CEEM, a leading American management system training and publication service provider, and International Standards Certification Pte Ltd, Singapore. In 2002, the ISO business of KPMG in North America was acquired. In 2003 BSI acquired BSI Pacific Ltd. for the greater China market. In the United Kingdom, BSI now holds 49% of the British Standards Publication Limited (BSPL).

In 2003 BSI celebrated the 100th anniversary of Kitemark, which is now a superbrand. In 2004 the KPMG certification business of the Netherlands was acquired by BSI, with its focus on Benelux and European business. In 2006 BSI acquired NIS ZERT of Germany (www.bsigroup.com/de-DE/, accessed 7 February 2016) Entropy International Limited, a software provider (www.bsi-entropy.com, accessed 7 February 2016) and Benchmark Certification Pty Ltd of Australia (www.benchmarking.com.au, accessed 7 February 2016).

In 2009 BSI acquired the supply chain division of First Advantage Cooperation in the United States, Certification International S.r.I. in Italy and EUROCAT for healthcare certification in Germany. In 2010 the BSI opened offices in Qatar and Saudi Arabia; in the United Kingdom the BSI acquired the gas certification body GLCS, in Italy it acquired BS Services Italia S.r.I. (BSS), and it also acquired System Management Indonesia (SMI) for Asian customers. This concludes a more than century-long path of growth for now and gives BSI a comprehensive business service portfolio.

The BSI is accepted and accredited by leading organizations like the ANSI-ASQ National Accreditation Board (http://anab.org/, accessed 2 March 2016), the China National Accreditation for Conformity Assessment (CAN), the Japan Accreditation Board (JAP) (http://www.jab.or.jp/en/, accessed 2 March 2016), the Raad voor Accreditatie (RvA) Service of the Netherlands (www.rva.nl, accessed 7 February 2016) and the United Kingdom Accreditation Service (UKAS) (www.ukas.com, accessed 7 February 2016).

5.6.2 National Standards Body

The UK government recognizes the BSI as a national standards body (NSB) formulated in the Memorandum of Understanding (MoU), first published 1981. This covers the management, coordination and undertaking of

- British standards – British standards are formal consensus standards and are based on the principles of European standardization;
- participation of the BSI in European and international standards bodies;

- promotion, marketing, distribution and information activities associated with British standards, BSI products and standardization generally;
- support and offer a corporate infrastructure for experts.

The director of standards has the primary responsibility for the activities set out above. The government and BSI agreed that:

- Standardization is a key factor for government policy including competitiveness, innovation, reduction of trade barriers, fair trading and protection of the customer's interests.
- Standardization is increasingly important for the globalization of commerce, convergence of technology and international trade following the World Trade Organization Agreement and European legislation harmonizing laws.
- Standardization is able to promote better legislation.
- Market forces are necessary for efficient standards but public policy is required to avoid impact on innovations.
- Standards can be used to reduce trade barriers and to promote innovation. Efficient standardization processes are needed.

International Role

The BSI plays an active role in the ISO, IEC and other relevant organizations to promote UK interests. The government works through the WTO TBT Committee to promote a coherent international standardization process and effective use of standardization for public policies.

In Europe, the BSI follows European standards policies. The government promotes the BSI as the national standardization body (NSB). The British Government recognizes the BSI as NSB for ISO, CEN, IEC, CENELEC and ETSI. The government supports the BSI in its efforts to harmonize European and international standards.

Commitments

The BSI upholds the public interest, provides information to the government, participates as a member of the major European and international standards organizations, fulfils EU Directive 98/34/EU on standardization, makes provisions for formal consensus standards and sets up a strategic plan to represent the United Kingdom's interests in standardization.

The government respects the independence of the BSI in standardization work, maintain a suitable environment for the work of the experts and provide financial support for the BSI to carry out all its national and international standardization duties.

5.6.3   Standard Development

There are more than 9000 UK expert volunteers active in standards. They come from manufacturers, users, research organizations, government departments and consumers. All new and revised standards are made available to the public for comments via the BSI draft review system. Input is accepted from any member of the general public who has an interest in the content of the standard.

Standards released by BSI are voluntary and not regulatory or imposed by government. They can be used to help support legislation or regulation. Being a BSI committee member is a substantial undertaking and for many people it is in addition to full-time employment. With registration in a committee the member has full access to all documents. Involvement in standardization requires reading documents circulated to committee members, nominating experts to carry out new work and attending meetings. Open-minded treatment is the basis of standardization work; conflicts of interest should be declared and confidentiality is maintained. National committee meetings take place at the BSI office in London.

It is the policy of BSI to publicize patents that apply to standards under development. Experts nominated by a stakeholder organization must keep close contact with their nominating organizations to make sure that they represent their interests.

The principle of consensus is the basis of standardization work and individual concerns must be carefully and fairly balanced against the wider public interest.

Documents are circulated for comments via the eCommittee. This reflects committee drafts of documents for comment, public comments and votes on documents and the final approval of the documents. For this process, standardized templates are used to collect the comments.

These comments are collated by the BSI and will be transferred to CEN, CENELEC, SIO, IEC, ITU or ETSI via the eCommittees, including votes on these documents. Delegates and experts must provide timely reports back to the UK committee when attending meetings in European or international working groups.

When developing a standard of UK origin, the first step is a proposal for a new work to be submitted to the BSI. Anyone can submit a proposal, which needs to be accepted by the BSI acceptance team. A draft will be circulated and approval of the draft will be made within 2 weeks by the BSI. The public comment period is 2 months and comments will be collected by interested parties through the draft review system. The comments will be considered by the related panel and decisions are made to accept the comments or reject them.

A final draft will be produced and circulated within 2 weeks. When consensus is reached the committee chairman and secretary will endorse the publication of the standard. Typesetting will be supported by the BSI production team. The published standard will be made available by the BSI shop within 12 to 15 months. A review of BSI standards will be initiated every 5 years.

5.7 USA – ANSI

5.7.1 General

The American National Standards Institute (ANSI) (www.ansi.org, accessed 7 February 2016) is the voice of the US standards and assessment system. It represents the United States as a national committee in the IEC, ISO and ITU. The mission of ANSI is explained as strengthening the United States in the global economy, ensuring the safety and health of consumers and protecting the environment. Table 5.30 gives an overview.

5.7.2 History

The United States is one of the early players in national and international standardization. As one of the founding members of the IEC in 1906, the United States started ANSI in 1919. It was originally established as the American Engineering Standards Committee (AESC), based

Table 5.30   ANSI.

Mission	Enhance global competitiveness of US business and US quality of life by promoting consensus standards and conformity assessment
Founded	19 October 1918
Legal status	Nonprofit organization
Locations	Headquarter in Washington DC
	Office in New York
Employees	90+
Membership	Governmental agencies, organizations, companies, academic and international bodies
Annual Budget	$36 million
Affiliations	International Organization for Standardization – ISO
	International Electrotechnical Commission – IEC
	International Accreditation Forum – IAF (http://www.iaf.nu/, accessed 2 March 2016)
Regional Affiliations	Pacific Area Standards Congress – PASC
	Pan-American Standards Commission – COPANT
	Pacific Accreditation Cooperation – PAC (www.apec-pac.org, accessed 7 February 2016)
	ANSI-ASQ National Accreditation Board (ANAB), a member of Inter-American Accreditation Cooperation (IAAC) (www.iaac.org.mx, accessed 7 February 2016)

Box 5.3   ANSI's founding organizations in 1919.

American Institute of Electrical Engineers (now IEEE)
American Society of Mechanical Engineers (ASME)
American Society and Civil Engineers (ASCE)
American Institute of Mining and Metallurgical Engineers (AIME)
American Society of Testing Materials (now ASTR International)
US Departments of War, Navy and Commerce

---

on an initiative of the American Institute of Electrical Engineers (now IEEE) in 1916. The goal was to form one national body to coordinate standards development and approve national consensus standards. The organizations that founded ANSI are shown in Box 5.3.

The organization's starting budget was $7500 annually. The first American Standard Safety Code was published in 1921 and covered the protection of the heads and eyes of industrial workers. In 1926 ANSI hosted the conference that created the International Standard Organization (ISA), which later became the International Organization for Standardization (ISO).

Connection to IEC

The International Electrotechnical Commission IEC was founded in 1906 in London. In 1931 the American Standards Association (ASA), a predecessor of ANSI, became the National Committee for the United States in the IEC. In the following years ASA grew to an internationally recognized organization of standards on safety in factories and households. It provided guidance on avoiding hazards.

The Second World War increased the standards activities of ASA, which became concerned with providing high-quality products for the military. More than 1300 engineers were working on standards for quality control, safety, photographical supplies and equipment components.

After the Second World War, the focus of standardization driven by ASA became more international. The ISO was founded together with 25 national standard bodies,. From the beginning ANSI played a strong, leading role in ISO and the IEC. In the 1950s and 1960s the new fields of standardization were nuclear power, information technology, material handling and electronics, which were provided by ASA in ISO and IEC.

In 1966 the ASA was renamed as the United States of American Standards Institute (USASI) and the first certification committee was founded in 1968. The present name of ANSI was introduced in 1969. In 1970 the ANSI Board of Standards Review (BSR) was founded and is responsible for the approval of standards. This is the most significant innovation of ANSI, which now sets guidelines on the standardization process of any consensus-based standard-writing organization. Today the ANSI observes more than 270 standard-writing organizations in the United States that are accredited by ANSI.

In the 1980s many activities were started to recognize the global view of business and to establish globally accepted standards as a key to unlock foreign markets. With the acceptance of the joint technical committees of the ISO and IEC, the JTC 1, a significant innovation in global standardization, was reached with the strong support of ANSI.

With the development of the European Union ANSI established a presence in Brussels, close to the EU Commission, the European Committee of Standards (CEN), the European Committee of Electrotechnical Standards (CENELEC) and the European Telecommunication Standards Institute (ETSI). In 1989 ANSI also started its outreach with offices in Eastern Europe, the Far East, the Pacific Region, South America and Central America. Discussions were started with Canada, Mexico and the United States on a North American Free Trade Agreement.

The 1990s brought a standardization focus on increasing global business. Companies not only saw the quality and safety aspect of standards for their products – they also used international standards to create global market access. Issues such as health and the environment played an increasing role in the daily standardization work. New standards have been created and existing standards have been extended, covering these topics.

In 2000 the first National Standards Strategy for the United States (NSS) was approved. The strategy gave a road map for standardization and defined a reliable, market-driven process.

In 2005 the new United States Standards Strategy (USSS), based on the NSS roadmap, brought new types of standard-setting activities into the picture, such as consortia and forums, to focus on more flexible approaches and new structures. The USSS is committed to following the World Trade Organization on the removal of trade barriers.

5.7.3 US Standardization System

The United States Standards Strategy and the National Conformity Assessment Principles for the United States support voluntary consensus standards and the related assessment programmes. Methods have been established for the US standardization system to develop standards and to follow the conformity assessment schemes for the stakeholder in the standardization process. The stakeholders in ANSI are more than 1000 members from industry, organizations and government agencies.

5.7.3.1 National Standardization

ANSI is the accreditor of standard-developing organizations (SDO), which develop American National Standards (ANS) in a voluntary national consensus standards process. Accreditation by ANSI proves that the standardization process of the SDO follows its main principles: openness to all interested persons or groups, a balance of interests, consensus-finding discussions and due process fixed in the operational procedures and bylaws of the SDOs.

In the United States ANSI has accredited more than 270 SDOs. The 20 largest produce more than 90% of all standards. In 2006 there were more than 10000 American National Standards published.

The assessment of SDOs by ANSI follows the ANSI Essential Requirements for a due process of American National Standards. Table 5.31 gives an overview of the ANSI Essential Requirements.

Table 5.31 ANSI essential requirements.

Essential requirements for due process	Due process means that any person (individual, company, organization, governmental) has the right to a. express a position b. have the position considered c. have the right to appeal. Due process allows for equity and fair process.
Openness	Participation of all individuals who are directly or materially affected.
Lack of dominance	The standard-developing process must not be dominated by the interests of any individual or organization.
Balance	Participants in the standard-developing process must be balanced, coming from different interest categories.
Coordination and harmonization	Potential conflicts between and among American National Standards must be resolved in good faith.
Notification	Notification of standards development must be given by suitable media.
Consideration of views and objections	Views and objections on a standard development must be considered promptly.
Consensus vote	Evidence of the consensus vote of the standard developer's accredited procedures must be documented.
Appeals	The standard developer (SDO) needs to have a written procedure for a readily available appeal mechanism.
Written procedures	The standard developer (SDO) needs to have written procedures for the standard-developing process available for any interested person.
Compliance	All ANSI-accredited standard developers (ASD) are required to follow the normative policies and administrative procedures of ANSI's executive standards council.

'Openness' requires that timely and adequate information should be published to publicize any activity associated with writing a new standard, revising an existing standard or withdrawing a standard.

A 'lack of dominance' in the standards process means that no person or entity should dominate the standardization process in any way.

'Balance' means that no single interest category should constitute more than one-third of the membership of a consensus body when safety requirements are involved or a single majority if no safety requirements are involved.

There are three categories of interest:

• producer;
• user;
• general interest.

There are several user categories:

• User – consumer. For consumer product standards.
• User – industrial. For industrial standards.
• User – governmental. For standards used by the government.
• User – Labor. For standards used in the interest of American workers

For notification and standards development coordination the Project Identification Notification System (PINS) has been established for American National Standards (ANS). The PINS covers all ANS and standards of ISO, IEC and ISO/IEC JTC 1.

Consideration of views and objections can be in written form to the standards developer or by commenting through PINS.

Evidence of consensus by the vote of the consensus body must be documented. The vote and comments must follow the criteria in Table 5.32.

The provision of appeals is important to protect directly and materially affected interests. At the standards developer level the right to appeal must be stated in the written procedures to offer an identifiable, realistic and readily available mechanism for the handling of appeals. Appeals can be made to ANSI. However, ANSI will not start its appeal process as long as the appeal process of the standard developer has not been concluded.

Table 5.32 Criteria for vote and comments of consensus standards.

Negative vote	Negative votes can only be changed to approve in agreement with the voter. Any negative vote must be addressed.
Records	All negative votes must be recorded and considered. Any change of negative vote needs to be recorded.
Negative vote without comment	These must be recorded and are not counted.
Voting	There are four ways to vote: – affirmative; – affirmative with comment; – negative with comment; – abstain.

ANSI Patent Policies

There is no objection in principle to drafting an ANS in terms that include an essential patent claim if the technical reason justifies this approach. An essential patent claim is given when the patent claim does not allow any other technical solution to fulfil the standard. The standards developer must receive a statement from the patent holder to cover the following:

- A general disclaimer that the party does not hold or intend to hold essential patent claims. (This is usually done by raising this question at the beginning of the work.)
- Assurance that a license for such essential patent claims will be made available to applicants wishing to use the license
  - under reasonable terms and conditions
  - without compensation.

This must be indicated in the standard in a note. This means that the user of the standard can receive a license to use the patent right for a reasonable and fair price.

Metric Policy

Each ANSI accredited standards developer must use the ANSI metric policy and use the International System of Units as preferred units for measurement in American national standards.

Publication

An ANS must be published at the latest 6 months after approval by the standard developer or ANSI.

National Adaptation of ISO and IEC

Standard developers that wish to adapt ISO or IEC standards must comply with the ANSI procedure for the national adaptation of ISO or IEC standards as American National Standards.

5.7.3.2   International Standardization

ANSI promotes the use of ANS on an international level in the ISO and IEC. It supports US policies and technical positions in international and regional standardization organizations. It encourages the adoption of international standards as national standards where they meet the needs of the user country.

ANSI is the only representative of the United State in the ISO and IEC. As a full member, the United States pays dues to both organizations and is an active member, with its citizens holding a large number of leading positions like chairmen and secretaries of technical committees and advisory boards. ANSI is recognized in the ISO and IEC as the US National Committee (US NC). It provides immediate access to ISO and IEC standard developing processes and participates in almost any of the technical committees with experts sent to working groups and maintenance teams.

Part of the responsibility of the US member body to ISO is to develop and transmit via ANSI the US positions on activities and ballot of the international technical committee formulated by the US Technical Advisory Group (US TAGs). The positions of the US in IEC are endorsed and closely monitored by the US NC Technical Management Committee (TMC).

In many cases US standards are taken forward to the ISO and IEC through ANSI or the US NC to be the basis for an ISO or IEC standard with few or no changes. In this respect ANSI plays an important role for US business, bringing US standards to an international level. The work in the international standardization groups of ISO and IEC is carried out by experts paid by industry not by ANSI staff. The success of this standardization work therefore depends on the willingness of industry to send experts.

5.7.3.3 Conformity Assessment

The term 'conformity assessment' describes the steps both by manufacturers and independent third parties to determine the fulfilment of standards' requirements.

ANSI's program of accrediting third-party product certification has experienced strong growth in recent years and will be in focus in the coming years as it strives to obtain world-wide acceptance of accredited certification performed in the United States.

The US government uses private-sector standards in regulations released by federal governmental agencies. State and local governments and agencies have formally adopted thousands of standards produced by ANSI and the process appears to be accelerating.

ANSI continues to be fully involved in the US and global standardization and conformity assessments.

5.7.4 Structure and Management

ANSI has more than 1000 businesses, professional societies and trade associations, standards developers, governmental agencies and consumer and labour organizations as members. They represent the diverse interests of more than 120 000 entities and 3.2 million professionals worldwide.

They are organized in four member forums as shown in Table 5.33.

The organization of ANSI is shown in Figure 5.16.

The members of the ANSI nominating committee are appointed to recommend candidates for seats on the ANSI board of directors and the board of leadership. The executive committee

Table 5.33   ANSI member forums.

Company Member Forum (CMF)	Forum of a broad spectrum of US industry to discuss national, regional and global standards and conformity assessment issues and collectively shape and influence US policy in the domestic and international arenas.
Government Member Forum (GMF)	A forum of government executives to discuss standards and conformity assessment issues as they relate to government agencies and their missions and objectives.
Organizational Member Forum (OMF)	A forum of US professional societies, trade associations, standard developers and academia to discuss national and international standards and conformity assessment issues.
Consumer Interest Forum (CIF)	Education of the consumer on activities of ANSI and the standard-developing community.

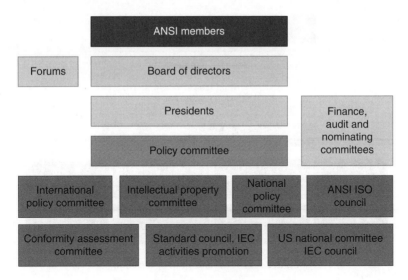

Figure 5.16 Organization of ANSI.

has the power to act for the board of directors between meetings of the board. The finance committee maintains a continuous review of the financial affairs of ANSI. The audit committee maintains a continuous review of the internal controls and audits affairs of ANSI.

5.8 Japan – JSA

5.8.1 General

The Japanese Standards Association (JSA) was founded on 6 December 1945 by merging the Dai Nihon Aerial Technology Association and the Japan Management Association through the Ministry of Trade and Industry. The office is presently located in Mita, Minato-ku and the objective of the JSA is to educate the public and industry regarding standardization. The JSA, together with experts from industry, supports national and international standardization. In this way it contributes to the improvement of technology and the enhancement of product efficiency. The JSA's organization is shown in Figure 5.17.

The JSA council elects the board of directors and they elect the president. The JSA's operational activities are carried out by the head office. Several units work on standard development, international standards, publication, training and certification. A standards council promotes the JSA's IEC activities.

5.8.2 Developing Standards

The JSA develops Japan Industry Standards (JIS) in various technical fields like graphical symbols, sampling inspections and quality assessment. The JSA has set up models and rules for the standardization process for drafting, commenting, voting and publishing. It is working on standards in basic technical and commercial fields like graphical image processing,

Figure 5.17 Organization of the JSA.

multimedia and information code exchange in the IT field. It also covers standards in environmental and electromagnetic fields. In cooperation with industry and other technical or governmental organizations, the JSA develops high technical standards for the Japanese market. The method used in Japan to develop standards is the Taguchi method for quality engineering.

The JSA carries out investigations and other studies on quality aspects of products and services in standards. For this reason, standards analysis methods are used to improve the quality and reliability of products and services.

Standards

The JSA conducts surveys and research in standardization of:

- basic and common fields such as units and graphical symbols;
- networking and software applications in the IT field;
- management systems;
- consumer protection.

It the JSA's responsibility to develop and maintain the high quality font of characters such as the Heisei Mincho typeface.

In the ISO and IEC, the JSA actively participates in the following technical committees:

- ISO TC 37 Terminology (principles and coordination)
- ISO TC 46 Information and documentation
- ISO TC 69 Application of statistical methods
- ISO TC 176 Quality management and quality assurance
- ISO TC 207 Environmental management
- IEC TC 1 Terminology
- IEC TC 3 Documentation and graphical symbols
- IEC TC 56 Dependability

These are some areas, among others, where the JSA is contributing very actively to international standardization in the ISO and IEC.

There are many Japanese experts active in various working groups of ISO and IEC.

5.8.3   Conformity Assessment

The JSA has established mirror committees on conformity assessment for the ISO Committee on Conformity Assessment (ISO/CASCO) and the IEC Conformity Assessment Board (IEC/CAB) to meet the obligations of the World Trade Organization on the removal of Trade Barriers (WTO/TBTR).

5.8.4   Education and Training

The JSA offers a wide range of seminars to improve education on standardization processes for industry and the public. These seminars provide information on how technical standards are written and how to use technical standards. This has a long tradition in Japan.

The first seminar on quality control was held in 1949 to provide information on management techniques to improve quality management and quality engineering. This seminar was held in major cities in Japan by experts from the JSA.

Today JSA provides seminars on ISO 9000 and ISO 14000 by internal auditors. The seminars include information on the Japanese JIS Mark Certification Scheme to bring quality control managers up to date on developments in quality management at an international level. A quality management (QM) and quality control (QC) examination is offered by JSA to educate industry experts for industrial standardization and quality management. For this reason the JSA offers education and training at companies in quality management, quality control and other quality techniques.

The JSA holds annual meetings in major cities in Japan as 'National Meetings on Standardization and Quality Management' to bring the topics of standards and quality to the attention of industrial experts, managers and public.

The issue of quality in Japan was brought into focus with the first 'Quality Month' in 1960. Since then the JSA has held education and training meetings on aspects of quality every month in major cities around the country. Textbooks and 'q(uality) flags' are issued and presented to successful participants in JSA seminars and training courses to highlight the importance of quality. The Union of Japanese Scientists and Engineers and the Japanese Chamber of Commerce are cosponsoring this. Figure 5.18 gives an overview of seminars given in Japan by the JSA to educate the public and industry experts. These seminars are given yearly or more frequently.

5.8.5   Publications

The JSA publishes and distributes the Japanese Industrial Standards (JIS). The JIS are major publications of JSA. For products or services that are important for the international market, JIS are published in English. For many products the ISO and IEC standards are used on the international market.

Quality management and control	ISO system
• Quality control and standardization • JIS quality control manager • QM and QC examination • Quality control • Improvement in work place • Quality control for concrete • QC circle • Design of experiments • Multivariate analysis method • Equipment management • Software application • QC by distance learning	• ISO 9000 • Auditor training • ISO 14000 • ISO/TS 16949 • ISO 22000 (Food safety management design) • ISO 27000

Quality engineering
• Quality engineering • Quality engineering by distance learning

Standardization/JIS
• Technical drawing engineering • Evaluation of uncertainties

Reliability
• Reliability

E-learning / distance learning	Customer satisfaction management
• E-learning • Distance learning	• Product planning • Policy management • Management system

Figure 5.18 Seminars on standards and quality by the JSA.

The JSA also publishes books on industrial standardization, quality management, administrative management, science and technology. Bestselling books are the *JIS Handbook, Glossary on Technical Terms in JIS,* the *Manual for Company-Wide Standardization,* the *Manual for Quality Management,* the *Manual for World Standards,* the Management Engineering series and the Easy Understanding series on science and technology.

The JSA publication *Standardization and Quality Control* was first published in 1946 and is a monthly magazine specializing in industrial standardization and quality management techniques. Quality-related products published by JSA are the *Book of JIS Colors,* the *JIS Color Chart, Grey Scales* for assessing colour change, three-dimensional standards, the *Statistics Simulation Kit* and the *Twenty Face Die for Generating Random Numbers.*

The JSA produces and distributes international standards for the ISO and IEC, the national standards of ANSI, BSI and others, and standards from developers such as ASME and ASMT.

All standards can be viewed at the JSA office libraries at the JSA headquarters in Mita and in the JSA offices in Kansai and Nagoya.

5.8.6 Cooperation with International Standardization

The JSA has cooperated actively with the ISO and IEC since these organizations were established. It sends representatives to ISO and IEC to develop standards and active participates at technical committee level and with advisory groups.

The JSA provides support for travelling delegates. Technical experts who participate in the standard-writing process at working group or maintenance team level are sponsored by industry, which covers travel expenses.

In ISO and IEC some secretarial positions are held by the JSA. Two examples are ISO TC 164 *Mechanical testing of metals* and ISO TEC 201 *Surface chemical analyses.*

In cooperation with developing countries the JSA provides assistance in standardization and quality management. This includes surveys, seminars and drafting development plans and it accepts trainees at JSA. These activities are commissioning by the Ministry of Economy, Trade and Industry (METI), the Japan International Cooperation Agency (JICA) and others.

Information on activities in the IEC is provided by the JSA International Standardization Forum and the IEC Activities Promotion Committee (IEC ARC). The JSA publishes the *International Standards Information* for the ISO and the *IEC APC News.*

5.8.7 Certification

Three certified management systems are established in Japan – for quality, environment and safety.

The quality management system was accredited as the JSA Management Systems Enhancement Department (JSA MSE Dept) by the Japanese Accreditation Board for Conformity Assessment in 1994. The JSA MSE Dept is now auditing and registration quality management systems applying ISO 9001.

The JSA MSE Dept supports Japanese industry in becoming more competitive in a global market business and in building trust in the high quality level of products and services of certified companies.

The Environmental Management System has been established at the JSA MSE Dept in 1997 as a certification body following ISO 14001.

The JSA MSE Dept can provide certification for the complete ISO 14000 system to help companies to fulfil international environmental requirements on management needs.

The Institute for Promotion of Digital Economy and Community (JIPDEC) received accreditation from the Information Security Management Systems for Japan in 2002. This new accreditation body provides certification under the Information Security System (ISMS) by applying ISO 27000. The intent of JIPDEC is to support industry in improving its management of confidentiality, integrity and availability of information that should be protected.

The Japanese Registration for Certificated Auditors (JRCA) was accredited by JAB as certification body for quality management systems in 1996. The quality management system had 9541 QMS auditors in 2010.

In Japan, there has been a growing trend for quality audits in industry since they started in 2005. This can be seen in the number of applicants that have fulfilled the requirements of a quality audit. In 2007 there were 20 000 successful audits. In 2009 the number increased to 80 000 and in 2010 a total of 120 000 successful audits were registered.

6

Standardization Support Organizations

6.1 General

There are several organizations whose goal is to prepare information to support standardization work. Some have an international character; others are regional or national in orientation; some are based on industrial-sector organizations. Support organizations for standardization prepare technical information from related technological fields. They consider how the technology in these fields has been applied and how it might be applied in future. In some cases user experiences with particular technology are collected and evaluated to gain an idea of its reliability. There may be investigations into the need for new technical designs to solve problems in practice. There are hundreds of such organizations around the world, combining thousands of experts to cover all technical fields. Only a few can be presented here.

6.2 CIGRE

The International Council of Large Electric Systems (CIGRE) has its headquarter in Paris. Founded in 1921, it is an international nonprofit organization with experts from more than 50 countries around the world. More than 3500 experts sponsored and financed by industry contribute to 16 study committees and their technical work is published in technical brochures or in the magazine *Electra*. The study committees are listed in Table 6.1.

Working Scope

The technical scope of the CIGRE study committees covers, in the A group, all kind of electric equipment needed for electric power generation, transmission and distribution. In the A1 Rotating Electric Machines committee, all types of electric generators and motors are covered,

Practical Guide to International Standardization for Electrical Engineers: Impact on Smart Grid and e-Mobility Markets, First Edition. Hermann J. Koch.
© 2016 John Wiley & Sons, Ltd. Published 2016 by John Wiley & Sons, Ltd.

Table 6.1　CIGRE study committees.

SC A1	Rotating Electrical Machines
SC A2	Transformers
SC A3	High Voltage Equipment
SC B1	Insulated Cables
SC B2	Overhead Lines
SC B3	Substations
SC B4	HVDC and Power Electronics
SC B5	Protection and Automation
SC C1	System Development and Economics
SC C2	System Operation and Control
SC C3	System Environmental Performance
SC C4	System Technical Performance
SC C5	Electricity Markets and Regulation
SC C6	Distribution Systems and Dispersed Generation
SC D1	Materials and Emerging Test Techniques
SC D2	Information Systems and Telecommunication

for large generators, renewable generation, machine monitoring and diagnosis – for example management and efficiency of electric machines.

In the A2 transformers committee, the design, application, maintenance, and operation of all types of transformers for distribution voltages (1 kV up to 52 kV), and transmission (52 kV up to 1000 kV) and DC converter transformers are covered by technical brochures. The A3 High Voltage Equipment Study committee covers all types of equipment like circuit breakers, switches and all types of physical phenomena related to switching.

The B group in CIGRE is related to assemblies of equipment like insulated cables, substations, high-voltage DC and protection and control. The B1 Insulated Cable Study committee covers power cables for laying in the soil as underground cables, or in a tunnel or duct, at sea as submarine cables and for high-voltage AC and DC applications. This includes the cable joints and the cable sealing ends and all thermal, mechanical and electrical aspects of the cable laying and its operation.

The B2 Study Committee for Overhead Lines is mainly involved in studying how public acceptance of overhead lines can be improved, how the transmission capabilities can be increased or how the reliability and availability can be increased.

The B3 Study Committee for Substations covers topics related to any aspect of high-voltage substations for switching and transforming in the distribution and transmission network. Life-cycle management, new substation concepts, substation management, maintenance, operation and the impact of digital communication and smart grids are topics of current interest.

The B4 Study Committee for High Voltage and Power Electronics covers aspects on converter stations and power electronic equipment. Requirements and technical developments are in the scope of this committee.

The B5 Study Committee for Protection and Control cover all aspects to protect and control high voltage substation and high voltage systems.

The C1 Study Committee for System Development and Economics covers aspects on developing the high voltage power system and the economical aspects on new technical developments.

The C2 Study Committee on System and Control covers the aspects which are related to the high voltage system and its control functions to keep the network stable.

The C3 Study Committee on Environmental Performance cover aspect on environmental impact of high voltage equipment and systems.

The C4 Study Committee on System Performance covers topics to improve the high voltage system in its function to deliver electric power with high reliability.

SC C5 investigates the consequences of regulatory changes for the electric power sector and the impact on the network. Market design driven by new generation technology (e.g. distributed renewable generation) or integration of intermittent generation and their impact on changes of the power transmission system.

The focus of SC C6 is on electric power distribution systems and the impact of dispersed generation like photovoltaic and wind energy, the existence of local independent network islands (Micro Grid) and active distribution networks reacting to demand management and active customer integration.

The D group in CIGRE is related to materials, testing and information systems. SC D1 covers the basic information on electrical insulation and conductor materials and high-voltage and current testing and measuring techniques. In D2 the work is focused on information and communication technologies.

Cooperation with Standardization

CIGRE as an international organization is officially linked to IEC and IEEE for the support of standardization work. The long list of liaisons between CIGRE and IEC in Table 6.2 shows the close coordination of both organizations. In the IEC, liaison type A is for international organizations like CIGRE and the links are made directly with the IEC technical committees with the relevant CIGRE study committee.

The different types of liaison are as follows:

- *Liaison type A.* This is seen as an organization that makes effective contributions to the work of IEC working groups and maintenance committees of the technical committees and subcommittees. It has full access to the documents and may nominate experts to the working groups and maintenance teams. It must be an organization with an international character or be broadly based in regional organizations. The liaison is established by the chief executive officer of the IEC and will be revised every 2 years.
- *Liaison type B.* This is for an organization that has indicated an interest in a particular technical field and wishes to be informed on the progress of work. For this it has access to the reports of the technical committees and subcommittees. It must be an organization with an international character or be broadly based in regional organizations. The liaison is established by the chief executive officer of the IEC and will be revised every 2 years.
- *Liaison type D.* Liaison type D is for organizations that want to make technical contributions to the working groups or maintenance teams. They include manufacturer organizations, commercial associations, industrial consortiums, user groups or professional or scientific societies with a multinational perspective. They must have sufficient representation in the related technical field. Type D liaison must be approved by the secretary of the technical committee or subcommittee, clearly indicating with which working group or maintenance team the liaison is linked. The rationale for the liaison must be clearly stated. The liaisons are reviewed every 2 years. Type D liaison organizations have the right to participate as full

Table 6.2 Liaison types A and B between IEC and CIGRE.

CIGRE	IEC
CIGRE	TC 1 Terminology
General secretary	TC 11 Overhead Lines
	TC 20 Electric Cables
	SC 22 F Power Electronics for Electrical Transmission and Distribution Systems
	TC 28 Insulation Co-ordination
	TC 73 Short-circuit Currents
	CISPR International Special Committee on Radio Frequency
SC A2 Transformers	TC 4 Hydraulic Turbines
	TC 10 Fluids for Electrotechnical Applications
	TC 14 Power Transformers
	TC 37 Surge Arresters
SC A3 High Voltage Equipment	SC 17A High Voltage Switchgear and Controlgear - Equipment
	TC 28 Insulation Co-ordination
	TC 37 Surge Arresters
	TC 38 Instrument Transformers
	Liaison Type B
	TC 36 Insulators
SC B2 Overhead Line	TC 7 Overhead Electrical Conductor
	TC 11 Overhead Lines
	Liaison Type B
	TC 36 Insulators
SC B3 Substations	SC 17C High Voltage Switchgear and Controlgear - Assemblies
	TC 38 Instrument Transformers
	TC 36 Insulators
SC B4 HVDC and Power Electronics	TC 22 F Power Electronics
	TC 115 HVDC above 100 kV
SC B5 Protection and Automation	TC 38 Instrument Transformers
	TC 95 Measuring Relay and Protection Equipment
SC C4 System Technical Performance	TC 28 Insulation Co-ordination
	TC 37 Surge Arresters
	TC 77 Electromagnetic Compatibility
	TC 77A EMC-Low Frequency Phenomena
	TC 77C High Power Transient Phenomena
	TC 95 Measuring Relay and Protection Equipment
SC D1 Materials and Emerging Test Techniques	TC 10 Fluids for Electrotechnical Applications
	TC 14 Power Transformers
	TC 42 High Voltage and High Current Test Technique
	TC 90 Superconductivity
	TC 112 Evaluation and Qualification of Electrical Insulating Materials and Systems
SC D2 Information Systems and Telecommunication	TC 57 Power Systems Management
	TC 95 Measuring and Protection Equipment

Table 6.3 Liaisons between IEEE and CIGRE.

CIGRE	IEEE
SC A2 Transformers	PES – Transformer Committee
SC A3 High Voltage Equipment	PES – Switchgear Committee
SC C2 System Operation and Control	PES – Power Systems Stability Controls
SC B3 Substations	PES – Wind Power Coordination Committee
SC B4 HVDC and Power Electronics	PES – Wind Power Coordination Committee
SC B5 Protection and Automation	PES – Wind Power Coordination Committee
SC C4 System Technical Performance	PES – System Technical Performance
SC C6 Distributed Systems and Dispersed Generation	PES – Wind and Solar Power Coordination Committee

members in the WG or maintenance team meetings. Type D liaison experts act as official representatives of the organizations that appoint them.

Questions are arising from the IEC standardization working groups on technical details to be clarified for standardization. This special work will be given to experts in a CIGRE working group to prepare technical knowledge and information as a discussion basis for the integration into a standard. The time schedule for a new IEC standard or a revision is too tight for detailed technical investigations so this work is delegated to CIGRE to prepare information for the next revision of this standard.

On the other hand, the technical work of CIGRE may bring information collected in a technical brochure, which is then the basis for new IEC standardization work. Both ways are often used and support the standardization work.

The same situation of supporting the standardization work of IEEE-SA and CIGRE with many links from CIGRE study committees to IEEE technical committees is shown in Table 6.3.

The cooperation of CIGRE with the IEEE Power and Energy Society (PES) is based on a contract called the Policy for Cooperation.

In many cases the coordination of the technical work between the standardization organizations of IEC and IEEE with CIGRE is linked by the same expert being active on both sides to make sure that the information can flow.

6.3 CIRED

CIRED Congrès International des Réseaux Electriques de Distribution (in English, the International Conference on Electricity Distribution) is a non-profit organization under Belgian law and has its headquarters in Belgium. The aim of CIRED is to increase business-relevant competences, skills and knowledge related to electricity distribution.

CIRED is organizing conferences for discussion and the exchange of information regarding new technical developments and experience in electric power distribution. In 1967 there was a conference on distribution in Edinburgh UK (ERA and IEE) and 1969 in Liège Belgium (AIM). Based on this experience there was the foundation of CIRED based in Belgium. The organizer of the conference which is held every second year alternating to CIGRE is rotational the AIM and the IET. The first conference of CIRED was held in May 1971 by the Association des Ingénieurs de Montefiore (AIM), Belgium. Support comes from the UK Electrical Research Association (ERA) and the Institution of Electrical Engineers (IEE).

CIRED covers topics of relevance to the electricity distribution community and is a place for the presentation and discussion of technical aspects. It covers the field of electricity distribution systems and associated services and equipment. It covers electricity supply topics including dispersed generation, renewable energy, smart grid, environmental impact, cost-reduction programmes business model related topics, regulatory aspects, customer services and technologies and the skills of experts in the distribution field, and educational programmes.

The technical conferences of CIRED is structured into six sessions:

- session 1: network components;
- session 2: power quality and electromagnetic compatibility;
- session 3: network operation, control and protection;
- session 4: distributed energy resources and active demand integration;
- session 5: planning of power distribution systems;
- session 6: challenges of DSO regulations and the competitive market.

CIRED is governed by the directing committee (DC), the advisory committee (AC) and the technical committee of supporting countries (TC), which are supported by national committees (NC) and the liaison committees (LCs). The officers of CIRED are the chairman of DC and the chairman of the TC. The conference organization is in the hand of the organizing committee (OC) and the TC and the CIRED secretary. The TC is composed of the committee chairman, the session chairmen and their rapporteurs as well as a liaison officer for joint working groups.

The conferences are held in Europe on odd-numbered years, hosted by different countries. On even-numbered years, CIRED workshops are organized around Europe. Technical papers are presented at each conference and workshop and can be downloaded free from http://www.cired.net/publications-all/cired-main-conferences (accessed 4 March 2016).

Working groups focus on topics such as

- Smart Grid at the distribution level;
- smart secondary substations;
- technical and nontechnical losses;
- TSO/DSO interface;
- CIRED-CIGRE joint working groups.

The benefits of participating in CIRED include continuously updated information about technical innovations and trends. Contact with experts in the same technical area permits networking and provides information about new innovations and best practice. Each session has a session advisory group (SAG) which are open to experts nominated by national or liaison committees.

7

Case Studies

7.1 General

The following case studies will illustrate the processes surrounding technological developments and standardization. In each case, economic interest is driving the processes.

In many cases the parties involved show a common interest. Even direct competitors can cooperate to create a common technical view and produce standards by consensus. The reason is simple: a standard will create a market for products and services and if there is a market the chances of each manufacturer to do good business is there. Standards also make products and services more comparable.

On the other hand, finding consensus in the working group in a technical area is not an easy task and sometimes may not be possible.

7.2 Smart Grid

7.2.1 General

Smart Grid was invented in the 1990s. Depending on the point of view of the person describing Smart Grid, it was a washing machine that started washing at night when the price of electricity is low, or a high-voltage transmission system that could control its power flow, or a real-time network configuration system that could shift electric energy from renewable generation resources to load centres depending on the availability of wind or sun and the need for electric power, or a micro grid that used wind, solar, biomass, gas turbine, combined cycle gas generation together with controlled user load of heating, cooling, light, and storage as needed.

All this smart energy generation, consumption and network management is based on digital communication and the Internet. The acronym 'smart grid' stands for all of this and should

Practical Guide to International Standardization for Electrical Engineers: Impact on Smart Grid and e-Mobility Markets, First Edition. Hermann J. Koch.
© 2016 John Wiley & Sons, Ltd. Published 2016 by John Wiley & Sons, Ltd.

indicate that there will be basic change, a new electric energy world based on the Internet and any new technology coming up.

Over years of discussion, the idea evolved of writing standards to make all this possible and compatible with using devices from different manufacturers.

The harmonization of digital communication (technical data and information using standard software protocols and formats) between all the devices in the network was one of the key elements for Smart Grid. With a global standard a global market can be formed.

The public can motivate the writing of standards and push developments. This was done in the United States by several Smart Grid projects in the NIST (www.nist.gov, accessed 7 February 2016), NEMA (www.nema.org, accessed 7 February 2016) and the IEEE. In the EU, the Energy Commission carried out projects that supported the technical development and formed groups to think about standards – a process that needs time.

On the technology side it was clear that if a common communication standard were available the manufacturers could design and develop equipment for the management and control of a Smart Grid. These new applications, based on a standard, could be designed in different places by different manufacturers in a competitive environment and users could find solutions for their needs on an economical basis.

The availability of standards in time is a key issue to market the products and services. So standards and products must be available more or less at the same time. This means first drafts should be ready after 1 year and standards ready after 3 years, which is a challenging timeframe given the need to go through all the discussion and voting steps of a consensus standard process.

7.2.2   Fragmented Environment

Looking at Smart Grid's environment, the United States, Europe and Asia were at very different levels.

International

The IEC started activities at the strategic level with Strategic Group (SG) 3 to consult the IEC standards management board (SMB). At this strategic level SG 3 developed an IEC Smart Grid standardization roadmap, which guided the IEC technical committees (TC) for its work on standards. The standardization is linked to working groups of the TCs as listed in Table 7.1.

Table 7.1   Smart Grid on an international level.

Strategy	IEC-SMB	Standards Management Board
	IEC-SG 3	Strategic Advisory Group
Guidelines	IEC-TC 57	Working Group 19 Task Force Smart Grid
Standards	IEC-TC 8	System Aspect
	IEC-TC 57	Energy Automation
	IEC-TC 64	Low Voltage Installations
	IEC-TC 65	Industrial Installations
	IEC-TC 82	Solar
	IEC-TC 88	Wind
	ISO-TC 205	Building Automation

Guidelines on Smart Grid standardization were formulated by IEC TC 57 in Working Group 19, Task Force Smart Grid. Here the need for digital communication between the different devices in the network was explained together with the urgent need to standardize communication protocols.

Based on a standardized communication protocol, devices for many functions in the network can be developed and can provide the intelligent functions a Smart Grid will need. This first work of standardization ended in the series of IEC 61850 standards, which are the basis for digital communication worldwide today.

Beside TC 57 for Energy Automation, the experts of TC 8 for electric power quality worked on standards that define the quality aspects of the network in terms of voltage stability, frequency stability, harmonics and overvoltages in the network. This becomes important with an increasing share of renewable energy that is fluctuating with the availability of power from sunshine and wind. Low-voltage installations (IEC TC 64) and industrial installations (TC 65) have many more automated functions controlled by digital communication, including the Internet. Signals about the price of electricity will influence the use of electric energy at the consumer or industry level. Energy saving by better and smarter control of processes will use the opportunities made available by a Smart Grid.

The solar technology of IEC TC 82 and the wind technologies of IEC TC 88 use Smart Grid control to provide the required energy in the network. Smart control of the network's power flow at the distribution and transmission voltage level will guide the renewable energy to the consumers in an ever-changing network configuration following the pattern of generation areas and load areas. All this is controlled in real time by the Smart Grid.

In ISO the Building Technology Standards of TC 205 open up new functions controlling the light, climate, heating, cooling, security and living ambience digitally through Internet-based connections to all devices related to a building.

United States

In the United States the National Institute of Standards and Technology (NIST) started together with Electric Power Research Institute (EPRI) to work out a Smart Grid roadmap to obtain an overview of the various activities using digital communication to make the electric network more intelligent. In a second phase the private entity EnerNex (www.enernex.com, accessed 7 February 2016) worked out an action plan on priorities for standardization works. These activities at the strategy level have generated information that is used to provide a guideline for the Smart Grid implementation in the standards world.

See Table 7.2.

Information about Smart Grid has been collected in IEEE P2030-2011 [4] to give an overview of the different activities with a focus on the interoperability of energy technology and information, technology operation with the electric power system, end-use applications and load. The guide has three parts: Part 1 on electric sourced transportation infrastructure, Part 2 on the interoperability of energy storage systems integrated with the electric power infrastructure, and Part 3 on standard test procedures for electric energy storage equipment and systems for electric power systems applications.

These documents were worked out and published between 2011 and 2014. They provide guidelines for many standardization activities in the United States, which are typically started by many different organizations driven by existing or expected business in this

Table 7.2 Smart Grid in the United States.

Strategy	NIST	Smart Grid Roadmap
	EPRI	Smart Grid Roadmap
	EnerNex	Smart Grid Priorities
Guides	IEEE	P2030 - Guide to Smart Grid
Standards	NIST	Action plan to co-lead IEC standardization
	IEC	TC 57 WG 10, 14, 17
	IEEE	SC C21
	UCA, SAE	Open Smart Grid, Multispeak
	NEMA	SG Advisory Panel Smart Grid
		ANSI, ZigBee, EPRI

technical sector. The United States, as a leading industrial country in electronics, communication technologies and software, has a wide range of expertise, many experts and companies to provide active standardization work. So it is impossible, here, to give an overview of all activities in Smart Grid standardization in the United States. Only some examples can be given.

One important decision by the United States under the lead of the National Institute of Standards and Technology (NIST) was to cooperate closely with and co-lead the international standardization activities led by IEC TC 57 for digital communication as one of the base standards for Smart Grid. ANSI, the US national committee at the IEC, therefore nominated a large number of experts to the more than ten working groups in IEC on the new standard 61850 *Communication networks and systems for power utility automation*. This series of standards covers any aspect of connecting digital devices in power systems, including software protocols, parameters and testing.

Beside the US activities (www.ucaiug.org, accessed 7 February 2016) in the IEC, a wide range of standardization activities may be seen in IEEE SC C21, the coordination group for Smart Grid in IEEE, with many standardization activities in the Power and Energy Society (PES) for substations, switchgear, control and protection, transformers and other power devices and systems.

The international user group (UCA) does not write standards but offers an international platform to standard users to develop information for standardization work for real-time application requirements. Big international companies like Siemens, ABB, Alstom, GE and many others are supporting the US activities.

One big project in the United States is the interoperability of the digital communication standard 61850 of IEC TC 57 for practical experiences with devices and systems of different manufacturers. With presentations of the functionality and exchangeability of different brands at international conferences and exhibitions, the United States supports the acceptance of this new technology and provides new information for the standardization process. Several other organizations are doing similar activities in the United States, like SAE International (www.sae.org, accessed 7 February 2016), Open Smart Grid (www.ucaiug.org, accessed 7 February 2016), Multispeak (www.multispeak.org, accessed 7 February 2016), a user forum for software compatibility, ZigBee (www.zigbee.org, accessed 7 February 2016), the open, global wireless standard to provide the foundation for the Internet of things, ANSI (www.ansi.org, accessed 7 February 2016) for standards organizations and EPRI (www.epri.com, accessed 7 February 2016), just to mention some.

Europe

Smart Grid activities in Europe show a very different picture from those in the United States. It is much more centralized towards the European Union and the related energy commissions, which required, in Mandate 441 of 2009, that CEN, CENELEC and ETSI must provide standards in the field of measuring instruments for the development of open architecture for utility meters involving communication protocols enabling interoperability. This mandate is very much focused on metering whereas in the United States Smart Grid activities started with a much wider scale of technical fields including digital communication and information.

The EU commission started investigations into Smart Grid in a task force to define the functionalities of meters and to set up regulations for data safety – an important consideration in Europe. See Table 7.3.

Guidelines in Europe are developed by the Task Force European Level Foundation (www. europarl.europa.eu, accessed 7 February 2016) as a meeting point for stakeholders in the field of Smart Grid and the Smart Meter European Coordination Group (www.cenelec.eu/aboutcenelec/ whatwedo/technologysectors/smartmetering.html, accessed 7 February 2016). Both groups are open forums for industry to participate in the standardization process in CENELEC with the Task Force Electrical Vehicles in TC 205 Home and Building Electronic Systems, TC 210 Electromagnetic Compatibility, TC 57 Power System Management and Associated Information Exchange and TC 13 Electric Energy Measurement and Control, and in CEN with TC 247 Building Automation and TC 249 Metering and in ETSI the Smart Grid Champion Forum M2 M Group (http://www.artemis-ioe.eu/ioe_project.htm, accessed 2 March 2016). The work on Smart Grid standards is very centralized under the EU standardization organizations (CEN, CENELEC and ETSI).

In addition to these standardization organizations there are a wide range of European industrial organizations like ENTSO-E (www.entsoe.eu, accessed 7 February 2016) or T&D Europe (www.tdeurope.eu, accessed 7 February 2016), to mention two organizations from the electric energy sector.

Table 7.3   Smart grid in Europe.

Strategy	EU	Commission Mandate 441 Metering
		Task Force
		Smart Grid Meters and Data Safety
Guides	EU	Task Force European Level Foundation
		Stakeholder Meeting
		Smart Meter European Coordination Group
Standards	CENELEC	Task Force Electrical Vehicles TC 205, TC 13, TC 57, TC 210
	CEN	TC 247 Building Automation
		TC 249 Metering
	ETSI	Smart Grid Champion Forum M2 M Group

Table 7.4 Smart grid in China.

Strategy	SAC	Standard Association of China
		Mirror Committee to IEC SG 3 Smart Grid
Guides	CEEIA	China Electric Equipment Industry Association
Standards	CEEIA	TC 82 (Mirror to IEC TC 57)
	SAC	TC 426 (Mirror to ISO TC 205)

China

In China the Standardization Association of China (SAC) is orientated to IEC for its Smart Grid standardization activities and has installed a mirror committee for the IEC SG 3 Smart Grid. See Table 7.4.

Guidelines on Smart Grid have been developed and published by the China Electric Equipment Industry Association (CEEIA) (www.ceeia.com, accessed 7 February 2016), a mirror committee to IEC TC 57. This is in accordance with the principal strategy in China of following the international standardization activities in the ISO, IEC and ETSI. If China's standardization requirements are not met then China will look for other internationally recognized organizations like IEEE, as was the case for the standardization of ultra-high voltage (UHV) systems of 1 000 000 V and above. After the IEC did not support writing standards on UHV, China started standardization activities with the IEEE Standards Association and published its first documents after only 3 years of work in 2012.

The CEEIA has installed TC 82 as a mirror committee to IEC TC 57 and the SAC has installed TC 426 as a mirror committee to ISO TC 205. This allows China to closely follow international standardization in the ISO, IEC and ITU to keep the world market open for Chinese products and services.

Germany

Smart Grid standardization activities have been started in the DKE (https://www.dke.de/de/Seiten/Startseite.aspx, accessed 2 March 2016), the German organization for standardization, to create a German standardization roadmap on Smart Grid. The newly founded Centre of Excellence Smart Grid has been formed as a mirror committee to IEC SG 3 Smart Grid to maintain close coordination with international standardization. In cooperation with German industrial organizations like the BDI (www.bdi.eu, accessed 7 February 2016), the VDE (www.vde.com, accessed 7 February 2016), or the ZVEI (www.zvei.org, accessed 7 February 2016), participation in the standardization of Smart Grid is widespread in Germany. See Table 7.5.

The DKE Centre of Competence Smart Grid prepared information to publish guidelines on Smart Grid by the DKE Focus Group. These guides include load management, smart meters, distribution, transmission, building automation, photovoltaic and wind.

Standardization work in Germany on Smart Grid for the electrotechnical side is carried out by the DKE and for buildings by the DIN. Both are working in close cooperation at the international level with the IEC (www.din.de, accessed 7 February 2016) and the (ISO www.iso.de, accessed 7 February 2016) and perform their work with mirror committees as shown in Table 7.5.

Table 7.5 Smart grid in Germany.

Strategy	DKE	Centre of Competence Smart Grid (mirror committee to IEC SG 3 Smart Grid) supported by BDI, VDE, ZVEI, etc.
Guides	DKE	Focus Group Load management, smart meter, distribution, transmission, building automation, photovoltaic, wWind
Standards	DKE	K 952 mirror to IEC TC 57 K 261 mirror to IEC TC 8 K M441 mirror to IEC TC 13 K 221 mirror to IEC TC 64 K 513, K518 etc.
	DIN	NA HRS mirror to ISO TC 205

7.2.3   Agenda

The worldwide agenda for Smart Grid standardization started in 2009 in the United States with NIST activities on a roadmap, followed in the same year by the IEC and its Strategic Group SG 3 on Smart Grid for another standardization roadmap. In Europe, the focus at the beginning in 2009 was given by the EU Mandate 441 on smart metering. In 2009 Germany also started its Centre of Excellence investigation on the impact of Smart Grid support by a wide coalition of industry associations. In the same year and the following years reports were published by those groups. The results of each group were used to enlarge the view of Smart Grid to include more and more technical fields involved in this new development. Today the term 'Smart Grid' has been transformed to express the fact that much more than just the grid is involved in the changes that are coming with the digitalization of communication and the Internet of things (http://www.artemis-ioe.eu/ioe_project.htm, accessed 2 March 2016).

The activities discussed here do not cover all of the activities on the topic of Smart Grid, which has been affected by standardization in all developed and industrial countries. Standardization activities influence each other and organizations learn from each other, often by the participation of experts in different organizations like IEC, IEEE and CENELEC. The organizations that published the first documents on Smart Grid did have a big influence on how standardization developed.

Today the term 'Smart Grid' also covers so-called micro grids. A micro grid is an independent network of renewable and distributed power generation and power consumption in a small local area like a city or a part of a city or a village with some farms. Micro grids combine different electricity generators, electric storage and electricity loads of the connected consumers. The network control is set up in a way that it will operate as a stable self-depending electrical system. An electric link to the main electricity transmission and distribution network acts as a backup for the electric power supply if the power generation within the micro grid is not sufficient to serve the power supply of the consumers on the micro grid. These micro grids are being developed rapidly today in the United States, Europe and Asia. They offer an independent power supply at low cost. This is the driver!

Different models are being implemented and different technologies are used. The number of possible combinations is enormous. An example is a new large housing area erected at an old military location equipped with photovoltaic, wind generators and gas turbines using methanol from a biogas plant. Energy is stored in batteries and by producing hydrogen (H_2) when a surplus of wind and solar energy generation is available. The Internet is the coordination medium to link all the necessary information together to provide a stable and safe network. And it works!

Comparing the Smart Grid activities world-wide the following picture can be drawn. The IEC plays a leading role internationally and many countries of North America, Europe and Asia are active participants. For most products and services, standards are already available today. The implementation of Smart Grid equipment in the old industrial countries is on a slow move forward because there is already a high-quality electric power supply system installed and functioning reliably. In countries with a high level of infrastructural investment, like the Arab countries or China, the spread of Smart Grid equipment and systems was faster and more accepted because of the missing existing infrastructure.

Today, Smart Grid applications are coming more quickly in industrial countries because of the reliability of renewable energy at lower prices. Once the micro grid is in place and installed the operating costs are low and most of the primary costs for fuel are zero like for wind, sun and bio waste. There are good experiences of new installations in Europe and North America. One of the first successful autonomous electric power generation and operation micro grids has been installed in Wildpoldsried in Germany (www.roadmap2050.eu, accessed 7 February 2016). The large variety of energy sources and loads like bio mass, bio gas, district heating, photovoltaic, solar, combined heat and electricity generation, hydro power plant and geo-thermal power generation offer stable electricity generation.

7.2.4   Controversial Topics

Nevertheless, standardization is also a field of technical competition and controversial topics. Driven by different positions in industry regarding how technology should develop, what would be the best technical solution is a controversial matter. Experts from different countries or companies promote their ideas in the standardization organizations and send a number of experts to bring their ideas forward. In some cases the manufacturer's name is known so that one can identify where the opinion is coming from but sometimes hidden behind consulting companies or other sponsored experts.

To manage and handle these influences in a proper way is the task of the related standardization organization. The final decision on which ideas or technical solutions will be used is made by the market.

Smart Grid architecture for digital communication requires the exchange of information standards to handle the data. This led to the international work in IEC TC 57 on energy automation.

Demand response is the reaction of the energy consumer to the availability and price of electricity in an automated network. In the United States energy profiles are already installed and established, again using the Internet as the communication background. Europe seems to be more conservative with regard to such technical solutions.

Renewable energy sources (RES) and distributed energy sources (DER) are promoted in Germany by the Renewable Energy Law (EEG) (www.bmwi.de/EN/root.html Energiegesetz, accessed 7 February 2016) and many installations are based on IEC 61850 standards while in the United States the standard used is IEEE 1547 with a different concept.

In metering, the ANSI C12 standard is promoted in the United States but also in Asia, whereas the European Union uses IEC TC 13 metering standards.

In e-mobility, Chinese and US standards are established and are being promoted rather than the IEC and ISO standards on international level.

Standardization activities by NIST, ANSI or IEEE compete in the market places with IEC and ISO standards and the market decides which to follow and to use.

The United States can be seen as the leader in application of Smart Grid solutions and standardization using both ANSI standards and the international standards of IEC and ISO, whichever fits best. In Europe, CENELEC and CEN offer a very homogeneous market and provide some basic standards for Smart Grid with IEC 61850. In Asia, China plays a strong role in adapting its own Smart Grid standards linked to IEC and ISO and in some cases also with the IEEE or other ANSI standards. The speed of installing Smart Grid solutions is high in China.

7.2.5   Interoperability

A key for the success of Smart Grid applications is the interoperability of equipment and software from different manufacturers. No user will go into a large control system if the equipment and software will come only from one manufacturer. Dependency is too great and no key market can develop. This is one main target of standardization – to open markets by making equipment and software compatible with other manufacturers' solutions. For this a two-phase investigation has been started in NIST to bring interoperability to Smart Grid solutions. In phase 1 it has identified an initial set of existing consensus standards and developed a roadmap to fill the gaps where interoperability is missing. This was done in workshops in 2009 and published as NIST Interoperability Framework 1.0. In the second phase a Smart Grid Interoperability Panel (SGIP) was established as a public-private forum for governance of the ongoing efforts. In a third phase, the Conformity Framework, including the testing and certification, was done in 2010 and 2011.

The tight timeline of these activities and the openness of the participating standardization organizations show that there is a high business expectation behind Smart Grid. It is envisaged that, within a few decades, the electric energy network, including power generation, will move toward regenerative energies because the advantages of Smart Energy are expected low cost and sustainability for the future as explained in the European Low Carbon Roadmap 2050 (www.roadmap2050.eu, accessed 7 February 2016). These expectations are driving industry and standards are setting the business framework.

Unlike in the past, competition in standards is seen as a positive aspect of getting the best solution. Standardization organizations today are much more willing to cooperate than in the past. The IEEE, in the United States, has cooperation agreements with the IEC on an international level and with CENELEC or SAC on a regional level, just to mention some examples. The speed of changes and the cost of system developments drive large companies in the Smart Grid business like Siemens, ABB, Alstom or GE to cooperate.

The main reason is the limited number of available experts. It is not possible to have multiparallel standardization activities. The standardization work is a continuous development: it will produce more standards in future and will change many existing standards.

Smart Grid technology is used in a wide spectrum of smart installations interacting with the availability of low price electric energy. Table 7.6 gives an overview of Smart Grid standards today.

Table 7.6 Overview Smart Grid international standards.

Architecture	IEC 62357	Seamless Integration Reference Architecture
Communication	IEC 60876	Transport Protocols
	IEC 61850	Power Utility Automation
Data models	IEC 61970/61968	Common Information Model (CIM)
Market	IEC 62235	Market communication Using CIM
Distributed energy Resources (DER)	IEC 61850	Power Utility Automation Part 7
Renewable energy sources (RES)	IEC 61400	Communication for monitoring and control of wind power plants
Security	IEC 62351	Security of Smart Grid
Metering	IEC 61334	Device Language Message Specification
	IEC 62056	(www.dlms.com, accessed 7 February 2016) Companion Specification for Energy Meterin
Home and building	IEC 50090	Home and Building Electronic Systems (HBES)
	ZigBee	Internet of Things

7.2.6   Challenges

Smart Grid is a complex horizontal standardization that impacts a wide range of technologies. A lot of stakeholders – in information technology (IT), in equipment for the energy sector, manufacturers and solution providers, and also the consumer – will find a new energy world with variable prices and quality.

The speed of innovation varies throughout the world and is changing over the time. The regulatory framework and laws are changing the situation in different countries and regions. There is a strong investment in wind power in Texas and a strong investment in photovoltaic in Arizona. In Germany wind and photovoltaic are in strong growth. These different ways of developing renewable energy generation relate to different regulations and laws and the financial support available.

Standardization development is like a moving target governed by activities in the various standardization organizations. The standard rules are changing and open different business opportunities but are subject to the influence of politics, laws and regulations. Only political intervention can stop a development or speed it up. This affects the results of standardization. Overall, today Smart Grid standards are facing a global competitive situation.

7.3   E-Mobility

7.3.1   General

To prepare effective standardization for e-mobility it is necessary to synchronize research, development, regulations and legal activities at national and international levels. It is necessary that standardization follows the required schedule so that it is in place when needed. The automotive sector is highly international and, therefore, it is necessary to provide international standards for e-mobility.

E-mobility comprises a large number of technical areas, which follow ISO for all mechanical aspects and IEC for electrical aspects and safety.

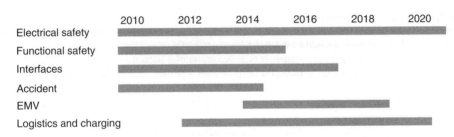

Figure 7.1   Overview of standardization activities.

The standards for e-mobility must be functionally orientated and not focused on technical solutions and performance based rather than descriptive. Interoperability for interfaces is required to provide solutions that allow different manufacturers to work together.

Worldwide coordinated battery-charging stations are needed. E-cars must be chargeable at any time at in any place and this process should be independent of e-car manufacturers and providers of infrastructure for charging batteries. Standards are needed to coordinate the unique charging process from plug to electricity bill.

In the automotive and electricity sectors, a large number of standards are already available to provide the right environment for e-mobility; they need to be coordinated and adapted. There are some technical fields that need new standards for unique plug and socket systems or interfaces.

E-mobility standardization must be driven mainly by international standards in the IEC and ISO. But standardization will also take place in regions like Europe, the United States or Asia with a strong automotive and electrical industry. Such regional standards must be transferred and harmonized on the international level to avoid specific technical solutions and to create a world market. An overview of standardization activities (https://www.dke.de/de/std/e-mobility/Seiten/E-Mobility.aspx, accessed 3 March 2007) is given in Figure 7.1.

7.3.2   Starting Point

Types of electric vehicles are divided into three main classes. Class M are vehicles with four wheels for passenger transport; class N are vehicles with four wheels for carriage of goods; and class L for two- or three-wheel vehicles and light quadricycles – see Table 7.7.

According to European Union Directives 2007/46/EG and 2002/24/EG, standardization today is focused on Class M electric vehicles for passenger transport with at least four wheels.

International-level standardization is carried out by the ISO, IEC and ITU. In Europe the equivalent CEN, CENELEC and ETSI organizations undertake the work and in the United States the International Consulting on Software and Interoperability (SAE) and the American National Standards Institute (ANSI) with the accredited standardization organizations like Underwriter Laboratories (UL) (www.ul.com, accessed 7 February 2016). Underwriter Laboratories develops full consensus-based standards.

In Germany the Road Vehicle Engineering Standards Committee of DIN (NA Automobile) is developing standards on electromobility and is supported by the German Automotive Industry Association (VDA) (https://www.vda.de/en, accessed 3 March 2016).

Table 7.7   Classes of electric vehicles.

M	Vehicles with at least four wheels for passengers
M1	Vehicles for personal transport of a maximum of 8 persons
M2	Vehicles for personal transport of more than 8 persons except the driver and a maximum weight of 5 tons
M3	Vehicles for personal transport of more than 8 persons except the driver and more than 5 tons weight
N	Vehicles with at least four wheels for carriage of goods
N1	Vehicles for carriage of goods up to a total weight of 3.5 tons
N2	Vehicles for carriage of goods above 3.5 tons up to 12 tons
N3	Vehicles for carriage of goods above 12 tons
L	Light vehicles with two or three wheels
L3e	Motorcycle with two wheels and a maximum speed of 45 km/h
L4e	Motorcycle with two wheels, a side car and a maximum speed of 45 km/h
L5e	Motorcycle with three wheels and a maximum speed of 45 km/h
L6e	Vehicle with four wheels, an empty weight up to 350 kg without a battery, a maximum speed of 45 km/h and maximum power of 4 kW
L7e	Vehicle with four wheels, an empty weight up to 400 kg without battery and a power of 15 kW

The standards from DIN are mirrored in ISO/TC 22 and CEN/TC 301 standards to avoid technical deviations on the international level with regional and national standards.

Vehicles in Europe have to follow EU regulations such as the Machinery Directive 2006/42/ EG released by the European Commission. To bring regulations to a global level, the United Nation (UN) releases Global Technical Regulations developed at an international level by the 'World Forum of Harmonization of Vehicle Regulation (WP 29)' of the United Economic Commission for Europe (UN/ECE). This global regulation is becoming increasingly important and its requirements will help to create a global automobile market.

Regulations for the transport of dangerous goods impose strong safety requirements, for example regarding lithium batteries, which pose a danger of fire, electric shock and even explosion.

7.3.3   Standardization Approach

The reason for the standardization of electromobility is to create a market for electric vehicles, road vehicle engineering, energy supply and associated information and communication technologies.

In Germany the government has installed the Nationale Plattform Elektromobilität (NPE), which has acted as a coordination group since 2010 and has released information about the standardization approach in its working group 4 'Standardization, Specification and Certification'.

This information takes an international perspective and is brought into the work of the ISO, IEC and ITU. There are several fields of standardization activities identified as shown in Figure 7.2.

The data flow for communication and the electric energy flow require standardized interfaces. Standardization needs include the software protocol for handling data in general,

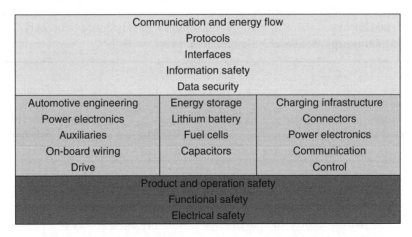

Figure 7.2 Standardization areas (https://www.dke.de/de/std/e-mobility/Seiten/E-Mobility.aspx, accessed 3 March 2016).

information exchange for safety reasons, and data security for electric vehicles. These are requirements for the safe, reliable and cost-effective use of an electric vehicle.

Automotive engineering focuses on the vehicle itself with electric power to drive the car and charge the battery, and any auxiliary component like speed indication, information about the status of the car, navigation and communication devices like smart phones that need to be integrated into the electric vehicle.

The storage system seen with electric vehicles includes Lit batteries, fuel cells and capacitors (MegCap) to provide the energy that the electric vehicle needs to operate.

The charging infrastructure includes connector technology, charging devices and communication and information technology to control, monitor and handle the charging process including the bill for electricity.

In standardization, the safety of the electric vehicle is seen from two sides: the functional safety in all cases of vehicle operation, including accidents, and the electrical safety of a highly concentrated high-voltage electric storage system.

There are different stakeholders around electromobility with the charging infrastructure, the vehicle and the battery as shown in Boxes 7.1 and 7.2.

There are a wide range of stakeholders in electromobility and they are mostly the same as the stakeholders in ordinary cars. The sales services, financing services as leasing, renting or credits, the traders of vehicles are already in place and only need adaptations. Testing, inspection, certification and billing of electricity are new requirements and standards are needed to harmonize these services on a worldwide basis but the principal structures are already found in today's car environment.

The use of digital communication with the Internet and integration of the smart phone into the electric vehicle operation are completely new and are done by Tesla (www.teslamotors.com, accessed 7 February 2016).

This environment requires new standards to define the performances of electric vehicles, the interfaces and to give explanations of terminology. Existing standards can be used and adapted for testing requirements and procedures. Finally, any of the products used in the electric vehicle should be standardized to guarantee compatibility.

Box 7.1   Stakeholders in electromobility (https://www.dke.de/de/std/e-mobility/
Seiten/E-Mobility.aspx, accessed 3 March 2016).

Services

Vehicle sales	Testing	Communication
Vehicle and battery financing	Inspection	Internet
Vehicle traders	Certification	Smart phone
Parking space management	Billing electricity	Arbitration

Box 7.2   Standards for innovation support (https://www.
dke.de/de/std/e-mobility/Seiten/E-Mobility.aspx, accessed
3 March 2016).

Standards

Performances	Test requirement
Interfaces	Test procedures
Terminology	Compatibility
Product standards	

The development speed of these standards is different in related areas. Some need to be released quickly; some have to wait.

7.3.4   International Agreements

The Emobility steering group was established in Germany (http://www.cencenelec.eu/standards/
Sectors/Transport/ElectricVehicles/Pages/default.aspx, accessed 3 March 2016). The aim of this group is the coordination of German activities in e-mobility and to bring these activities to an international (IEC) and European (CENELEC) level. This work is a common activity in the electric standardization of DKE and DIN/NA Automobil. The main task is to coordinate existing working groups and experts instead of creating new working groups. The coordination groups at the German level are shown in Table 7.8.

The DKE and DIN established the DIN electromobility office. This organization in Germany coordinates any standardization activity related to electromobility. Coordination involves setting up the standardization roadmap, coordinating standardization activities with China's SAC and bringing forward the battery standardization for storage.

On the electrical side, a series of standards is already in place covering charging with AC or DC, connectors, protective devices and EMS including radio interference. On the automotive side of DIN standard for vehicle to grid communication, electrical safety and energy storage are available.

There are some research activities to support the standardization process for battery applications and for the use of charging stations. For charging stations three different technologies

Table 7.8 German standardization (https://www.dke.de/de/std/e-mobility/Seiten/E-Mobility.aspx, accessed 3 March 2016).

Electric	Electromobility Office	Automotive NA Automobil
DKE	DIN	DIN
DKE/GAK 353.0.1 Inductive charging	Steering Group	NA-052-01-03-17 GAK
DKE/GAK 353.0.2 DC charging	EMOBILITY	Vehicle to grid
DKE/GAK 353.0.4 AC charging		
DKE/GAK 353.0.9 Energy supply	EMOBILITY	NA-052-01-21-01 GAK
DKE/GAK 541.3.6 Protective devices	Roadmap	Electrical safety
DKE/GAK 542.4.1 Connector system		
DKE/GAK 542.4.3 DC-connector system	EMOBILITY	NA-052-01-21-03 GAK
DKE/GAK 767.13.18 EMC	German-Chinese	Energy storage
DKE/GAK 767.14 Radio Interferences		
	EMOBILITY Batteries	

Table 7.9 European activities on electric vehicle standards (https://www.dke.de/de/std/e-mobility/Seiten/E-Mobility.aspx, accessed 3 March 2016).

Euro NCAP	Test Procedures for Safer Cars in Europe
US NCAP	US New Car Assessment Program
ETSI TC ITS	Intelligent Transport Systems Car to Car Communication
W3C	Worldwide web consortium on standards like XLM protocols

are used: data migration by mobile phones, radio frequency identification data (RFID) to identify physical devices and services and the logical characteristics and electronic codes with contract numbers and ID schemes via the Internet.

The German activities are integrated into the international activity in ISOL and IEC in joint working group ISO/TC22 SC3 WG1.

The link with the European level of German standardization of electric vehicles is through the EU Mandate 468: Mandate to CEN/CENELEC/ETSI for the charging stations of electric vehicles (http://ec.europa.eu/growth/tools-databases/mandates/index.cfm?fuseaction=refSearch. search&id=450, accessed 8 February 2016). The following activities have been started in EU on the electric vehicles, see Table 7.9.

International standardization activities on electromobility are organized at the European level with Euro NCAP (http://www.euroncap.com/en, accessed 3 March 2016), an organization which provides test procedures for safer cars in Europe. The US equivalent is US NCAP (www.safercar.gov, accessed 8 February 2016), which operates a New Car Assessment Program. On the international level, when standardization is organized by ETSI (http://www.etsi.org/technologies-clusters/technologies/intelligent-transport, accessed 8 February 2016), a technical committee is focused on intelligent transport systems with car-to-car communication. And the World Wide Web Consortium W3C (www.w3.org, accessed 8 February 2016), is publishing standards that are linked to the Internet and XLM protocols.

7.3.5 System Overview

Electric mobility is very closed linked to Smart Grid as explained above. The electric vehicle can be seen as an electric storage system. In normal use, the electric vehicle is used less than 10% of the time for the purpose of moving people around. In most cases the vehicle is not moving and waits for the next journey. A Smart Grid in the future will use the electric car and the battery in it to store energy (when the electricity price is low) or to feed back the electric energy to the power supply network (when the energy price is high). This could lead to a so-called virtual power plant when many cars are coordinated by the Internet, when necessary, to store electric energy from the network or to feed it back when needed.

Charging the Battery

The charging can be done at a private level (a private garage), quasi-public (a parking lot of a company allowing access) or public (provided by the city or a supermarket). This charging may be combined with parking. The location can be at the side of a street side or a parking space. Charging can be carried out with a private single-phase supply from a house. On the other hand, fast-charging stations will be needed, similar to today's gas stations, when electric vehicles are on long-distance journeys. All these different charging locations need a standardized charging process adapted to the charging situation and requirements.

Charging Functions

There are standard battery-charging connections of 16 A maximum charging current. Quick charging using AC and DC will need higher current ratings for the connections, then the charging can be line connected or it can use induced currents. There are control lines to control the process of charging the battery and the tariff system, through which the power supplier might limit charging to a particular maximum current or prevent charging at high loads during the day, particularly at midday. The load management of Smart Grid solutions will help to coordinate the charging of electric vehicles without overloading the power supply system; and the electric vehicle battery can be used to feedback energy from the vehicle battery to the power supply network. The connections of many car batteries through the Internet can provide energy in the range of a large power plant, at least for some minutes, to provide network stability.

The integration of electric vehicles into the operation of the power supply grid can be managed in different ways as shown in Table 7.10.

The issues surrounding electric vehicle power supply are charging, prices, loads and feedback between power supply grid and vehicle. The customer may choose a charging time and a charging profile – for example, battery charging after work with a maximum current of 16 A or during work at the company's parking space with 32 A for 4 hours. When prices are relevant to loading cycles, the user of an electric vehicle might charge at a low-price

Table 7.10 Integration of electric vehicles to the power supply grid.

	Charging	Prices	Loads	Feedback
Customer	Choose time and charging profile	Time-based tariffs	Desired loads	Desired load no feedback
Charging provider	No influence on charging	Indirect influence on charging time	Adapted loads	Control of loads

Table 7.11 Electromobility system overview (https://www.dke.de/de/std/e-mobility/Seiten/E-Mobility.aspx, 3 March 2016).

Services	Measurements and billing
	Diagnostics
	Load management
	Generation management
Communication	Media
	Protocols
	Signalling
Physical Level	Electrical parameters
	Mechanical parameters

session – for example during some off-peak sessions. The loads may be archived by the user or may be adapted by the charging provider to the availability of low-price energy and the limitation imposed by the maximum load of the supply lines. Then the Smart Grid can offer many ways to communicate with the electric vehicle to optimize charging based on a feedback process through the Internet. The electromobility can be divided into services, communication and physical levels as shown in Table 7.11.

The services associated with electromobility are covering the measurement of electrical energy used or generated and providing a bill for the used energy or a credit for delivered electrical energy. Diagnostics required for the operation and maintenance of the vehicle can be collected and provided for the maintenance process. The load management system provides information about the required electrical energy and the reserves in the battery.

A wide range of communication data can be made available by the different media installed in the communication system of the vehicle, including continuous contact with the Internet and an exchange of information about the vehicle's technical status, its location and the planned travelling routes. Information about failure signals, maintenance signals or position signals can provide information about repair requirements or route changes in real time. On the physical level electrical and mechanical data are provided on the status of the vehicle.

Communication between the vehicle and the charging station is standardized in ISO/IEC 15118 *Road vehicles* and ISO/IEC 14534-3 *IT-Home electronic systems*. Web-based services use the Payment Cards Industry Security Standards Council (PCISSC) (https://de.pcisecurity standards.org/minisite/en/, accessed 3 March 2016) and the EMV Co (https://www.emvco.com, accessed 8 February 2016) chip-based payment instrument with rules similar to credit cards.

Load Management

The load management and the Smart Grid can offer the user different levels of choice depending on different prices and tariffs from different power providers. Offers of renewable, CO_2-free electricity will also influence the user's behaviour when charging the battery. The load management of electric vehicles offers new possibilities to trade and store electric energy.

Storage Management

The electric vehicle battery in a Smart Grid environment provides an easy-to-use storage opportunity for electric energy on a short term notice. Loading the battery takes electric power from the network in cases if there is a surplus of wind or solar energy and on the reverse it can give electric energy back to the power grid if wind and solar energy is missing. Controlled through the Internet, the large capacities of electric vehicle batteries add up to a volume that is equivalent to that of a power plant. A typical e-car battery has about 60 kWh of energy storage capacity. One thousand such cars offer total energy storage of 60 MWh, this is the power of a middle size power plant. A large-scale power plant of 600 MW generates in 1 hour the equivalent energy of the storage capacity of 10 000 electric vehicles. Using the electric vehicle battery storage capacity of in a virtual connection controlled by the Internet makes it possible to store the electric power of a large-scale power plant in the electricity network for 1 hour. In Germany today about 80 million cars are in service, most of the time standing. This would give a backup energy storage of 4800 GWh, a large storage capacity. Even if only a part of this storage were available, it could contribute to stabilization of the power flow in the network.

Managed by an intelligent storage management using the Smart Grid data communication protocols, this trading of storage capacity is a service that produces profit when electric energy is stored when the price is low and sold to the network when the price is high. All this action can be automated by the storage-management software.

Data Security

With experiences of data safety on the Internet made in recent years the data security of an electric vehicle is a key element in the successful use of this technology. The control system of the vehicle can cause great danger when manipulated. Driving depends on secure data and manipulation needs to be avoided. Here standards are needed to secure the data in the vehicle and the network through the charging stations. Service centres for maintenance management also needs protection and data security to provide correct data for the billing system. ISO/IEC 27000 *Management systems of information security and digital communication* (www.27000.org, accessed 8 February 2016) of power networks are currently following the data-communication protocol of IEC 61850 for power-system automation (www.energy.siemens.com/hq/de/energy-themen//iec61850.htm, accessed 8 February 2016).

The data safety between the vehicle and the charging station is defined in IEC 61851-24 *Digital communication between a d.c. EV charging station and an electric vehicle for control of d.c. charging*, IEC 61851-23 *DC electric vehicle charging station* and ISO 15118 *Vehicle to grid communication interface of road vehicles*.

ISO 15408 gives requirements on information technology for evaluation criteria for IT security. This includes the protection protocols for digital travel data, metering interfaces and the Smart Grid connection interface. The data security around electric vehicles has a complex structure and some basic principles are shown in Box 7.3.

Box 7.3   Principles of data security.

Data sovereignty
Data avoidance
Pseudonymization
Data minimization
Grammar of transmitted data
Limitation of authorized data receiver or user
Manipulation protection
Data with personal reference

7.3.6 Electric Vehicles

As mentioned above, there are three classes of electric vehicles. The M class is for four-wheel cars to transport persons; the L class is for light vehicles with two, three or four wheels and the N class is for carriage transport. The electric vehicles here are only powered by electricity with batteries, fuel cells or capacitors.

Hybrid vehicles are also equipped with a combustion engine to charge the battery while the electric motors drive the vehicle.

Safety

There are three safety elements of electric vehicles. One is related to personal safety and the danger of electric shock with the high voltages used in the electric vehicle (60 V and 600 V).

According to ISO 6429 *Information technology – control functions for coded character sets* and ISO 6722 *Road vehicles – 60 V and 600 V single core cables – dimensions, test methods and requirements,* rules are different to fulfil electric safety requirements. The electric vehicle requires standards for crash situations to secure passengers and other individuals, to inform the rescue teams and to deal with the complexity of information exchange between any parties involved. This is part of the functional safety around electric vehicle architecture as defined in ISO 26262 *Road vehicles – functional safety,* which covers vocabulary, management of functional safety, the concept phase, product development at the hardware and software level, production and operation, automotive safety integrity level (ASIL) and supporting processes.

Battery

Today standardization focuses on lithium-ion batteries. The battery for the driving power takes up a substantial portion of the total weight and cost of an electric vehicle. The standardization of such a battery would limit technical development to future innovations. The weight of the lithium-ion battery, its functionality and the user friendly design will change with innovations and should not be limited by standards.

What must be standardized are the dimensions of connectors of single battery cells and their electric properties. In ISO 12405 *Electric propelled road vehicles,* definitions and requirements are given for high-power and high-energy applications and safety performance requirements and in IEC 62260 *Secondary lithium-ion cells for the propulsion of electric road vehicles* on performance testing and reliability. So the battery of an electric vehicle is seen as a highly innovative element where standardization can allow smart system solutions.

Fuel Cells

Fuel cells todays are not standardized as it is expected that they will be subject to technical development.

Capacitors

So-called supercaps or ultracaps are new capacitors using the so-called double-layer capacitor technology to reach high energy values for electric storage. In IEC 62576 *Electric double-layer capacitor for use in hybrid electric vehicles – test methods for electrical characteristics,* test procedures are defined to prove the quality of these capacitors.

Table 7.12 gives an overview of the international standardization of electric vehicles.

The overview in Table 7.12 gives an indication about the complexity of the standardization work related to electric vehicles. In close cooperation between ISO and IEC over a period of

Table 7.12 Overview of international standardization of electric vehicles.

ISO 62660	Secondary lithium-ion cells for propulsion of electric vehicles	2010
	Part 1: Performance testing	
	Part 2: Reliability and abuse testing	
ISO 6722	Road vehicles 60 V and 600 V single core cable	2011
	Part 1: Dimensions, test methods and requirements for copper conductor cables	
	Part 2: Dimensions, test methods and requirements for aluminium conductor cables	2013
ISO 6469	Electrically propelled road vehicles – Safety specification	
	Part 1: Onboard rechargeable energy storage system (RESS)	2009
	Part 2: Vehicle operational safety means and protection against failures	2009
	Part 3: Protection of persons against electric shock	2011
	Part 4: Post crash electrical safety	2015
ISO 8713	Electrically propelled road vehicles – Vocabulary	2012
ISO 11452	Road vehicles - Component test methods for electrical disturbance from narrow band radiated electro-magnetic energy	
	Part 1: General principles and terminology	2015
	Part 2: Absorber-lined shielded enclosure	2004
	Part 3: Transfer electromagnetic mode (TEM) cell	2001
	Part 4: Harness excitation methods	2011
	Part 5: Stripline	2002
	Part 7: Direct radio frequency (RF) power injection	2003
	Part 8: Immunity of electric fields	2015
	Part 9: Portable transmitters	2012
	Part 10: Immunity to conducted disturbances in the extended audio frequency range	2009
ISO 12405	Electrically propelled road vehicles – Test specification for lithium-ion battery packs and systems	
	Part 1: High-power applications	2011
	Part 2: High-energy applications	2012
	Part 3: Safety performance requirements	2014
ISO 14572	Road vehicles	2011
	– Round, sheathed, 60 V and 600 V, screened and unscreened single- or multi-core cables	
	– Test methods and requirements for basic- and high performance cables	
ISO/IEC 15118	Road vehicles – vehicle to grid communication interface	
	Part 1: General information and used-case definition	2013
	Part 2: Network and application protocol of requirement	2014
	Part 3: Physical and data link layer requirement	2015
ISO 23274	Hybrid-electric road vehicle – exhaust emissions and fuel consumption measurement	
	Part 1: Non-externally chargeable vehicles	2013
	Part 2: Externally chargeable vehicles	2012
ISO 26262	Road vehicles – function safety	2011
	Part 1: Vocabulary	
	Part 2: Management and functional safety	
	Part 3: Concept phase	
	Part 4: Product development at the system level	
	Part 5: Product development at the hardware level	
	Part 6: Product development at the software level	
	Part 7: Production and operation	
	Part 8: Supporting processes	
	Part 9: Automotive safety integrity level (ASIC)	
	Part 10: Guideline to ISO 26262	

about five years from 2010 to 2015 most of the standardization work has been started or concluded. In a full consensus standard, which is the goal for ISO and IEC, this each single step requires consensus, starting with the approval of the proposal for new standardization work, the first draft document for comments, the draft document for vote and the final draft for vote. At the same time the standardization of such a new innovative technology must not hinder or block technical development by fixing design requirements in standards. That is why the standardization is more focused on the functionality, not on the specific technical solution. A technical solution may change quickly; to change a standard needs 3 to 4 years, which is too long for innovative technologies. The electric vehicle today is, from a standardization point of view, well prepared for a successful and innovative future.

7.3.7   Charging Station

An electric vehicle needs to be charged. This charging is comparable to filling the tank of a combustion vehicle with gasoline but it usually needs more time and is required after shorter periods.

The technology used in the charging station – single and three-phase AC voltages or DC voltage – has a great impact on the charging speed and also must fit the needs of lithium-ion battery because incorrect charging may destroy the battery, which is expensive. In addition to the electrical conditions that affect battery charging, environmental conditions such as the ambient temperature are also of great importance. So the charging station has to fulfil several requirements as shown in Box 7.4.

The safety requirement incudes the safe connection of the charging station to the network and to the vehicle. The network load requirement of a charging station is different from that which we are using today. Charging an electric vehicle battery requires loading times at the rated power level for hours. Most private installations are not made for this type of application so there is a danger of overheating and fire.

The load current depends on the ambient temperature. At cold temperatures around and below $0\,°C$, the risk of damaging the lithium-ion battery is high. On the other hand, when ambient temperatures are high, for example $40\,°C$, the additional charging current might be too much and the battery might be damaged too. So the charging station has to control the charging process to be on the safe side.

Box 7.4   System requirements to the charging station.

Safety
Wide availability
Charging time
Comfort of charging process
Equipment need on the vehicle
Load management
Energy storage and recharge to grid
International compatibility

It should be possible to charge an electric vehicle at many locations. An electric vehicle with an empty battery is not very helpful. Therefore, different charging possibilities are recommended. From my own experiences in the city of Berlin, the fixed charging stations around the city were helpful and allow a fast-charging process using three-phase AC connections. In some cases it is also possible to charge the battery with a standard single-phase cable.

The charging time depends on the use of the vehicle. As the vehicle is standing still more than 90% of the time there is much time for charging; but when it is on the road travelling, the charging time must be short – less than half an hour – which is possible even with today's technology.

The complex technical requirements involved in charging, which depend on factors such as temperature and available network connection points, must be handled by the charging station. Once the driver has connected the plug, the only information that he should need is when the battery is charged and he can continue his trip.

The equipment needed in the vehicle provides independence if all the electronic charging devices are built into the vehicle but this has a cost in terms of additional weight. If the electronics are in the charging device, then the vehicle is lighter but it needs specific charging stations. This is an operational decision for the user of the electric vehicle.

Load management systems in electric vehicles can provide special tariffs for the network operator to prevent overloading the network, for example if all the electric vehicle drivers were to come home at 17:00 and fully charge their batteries. Without load management this would cause an overload – the network is not planned in that way. A load-management system would time shift the charging process.

The same applies when using the battery of the electric vehicle as energy storage for the network. The system should store electrical energy when too much energy is available in the network or feedback energy when the generation in the network is low. So many batteries can be used as a virtual power plant to stabilize the network. For the electric vehicle user, this network stabilization will bring additional income for the service.

The rules to operate electric vehicles must be international and under the same standards. This will create a global market and low prices for users of electric vehicles.

Charging Modes

There are four charging modes. The charging mode 1 is shown in Box 7.5.

Mode 1 is the simplest charging mode, with a maximum charging current of 16 A in a 240 V (AC) single or 480 V (AC) three-phase system. There is no safety device in the charging cable

---

Box 7.5   Charging mode 1.

AC charging current 16 A
250 V (AC) single or 480 V (AC) three phase
No safety device in the charging cable
Residual current device (RCD) in the house installation
No energy recharging to the network
No communication
Not accepted in some countries, e.g. United States

from the battery-charging electronics to the power supply of the network; the residual current device (RCD) for fault currents is in the house installation. Recharging from the vehicle battery to the network is not possible and there is no communication between the electric vehicle and the charging connection to the network. This single charging mode is not allowed in some countries, such as the United States.

Charging mode 2 is shown in Box 7.6.

Charging mode 2 allows a 32 A (AC) current for charging the battery at 250 V (AC) single phase or a 480 V (AC) three-phase network connection. This provides double the energy of mode 1 and can cut the charging time by half. The charging cable is protected by the so-called 'in-cable control box', which provides a residual control device (RCD) and a control pilot to control charging voltages and currents.

There is no option to recharge the battery of the vehicle to the network. Communication is given between the in-cable control box, the electric vehicle and the control pilot of the charging station.

Charging mode 3 is shown in Box 7.7.

Charging mode 3 offers an AC charging current of 63 A at specially installed charging stations. Normal house installations cannot be used. The high current allows charging times that are about half of mode 2 and one tenth of mode 1. The charging station can be connected to 250 V (AC) single phase or 480 V (AC) three phase and the plug is in accordance with to IEC 62196-2 *Plugs and sockets*.

No in-cable control box is required because the charging station provides safety devices. The interlockability of the plug/socket system allows the electric vehicle to be charged at public charging stations. Recharging to the electric supply network is possible with bidirectional communication, control and plug/socket interlocking.

Charging mode 4 is shown in Box 7.8.

Box 7.6   Charging mode 2.

AC charging current up to 32 A
250 V (AC) single or 480 V (AC) three phase
'in-cable control box'
No energy re-charging to the network
Communication between 'in-cable control box' and electric vehicle

---

Box 7.7   Charging mode 3.

AC charging current up to 63 A
250 V (AC) single phase or 480 V (AC) three phase
Charging plug according to IEC 62192-2
No 'in-cable control box' required, safety devices are part of charging station
Interlocking of plug/socket
Recharging to power-supply network possible

> **Box 7.8**   Charging mode 4.
>
> ---
>
> DC current charging station
> DC voltage and current system dependent
> Charging cable included control
> Complex safety functions for DC
>
> ---

Mode 4 uses DC current to charge the battery directly. This is the most powerful and fastest way to charge the battery. The currents and voltages are system dependent today and standardization is needed. The charging process is fully controlled in both ways to store energy in the vehicle battery or to recharge the battery to deliver electric energy to the connected power network. This can change in seconds.

It is valuable for network stability in renewable generation environments. Many electric vehicles can be interconnected to large virtual power plants distributed over the network. The DC charging stations are under innovative development and need more standardization.

Inductive Charging

The latest trend in charging electric vehicle batteries is by using inductive couplers. In this case the electric vehicle is connected to the charging station by inductive couplers (coils). The first IEC standard was published as IEC 61980-1 *Electric vehicle wireless power transfer (WPT)*. The inductive charging process is still under development.

Plugs and Sockets

Three types of plugs have been standardized in IEC 62196, see Table 7.13.

The standardization process for the plug-and-socket system of the electric vehicle is an example of real international cooperation and technical discussion reaching a worldwide consensus. Given the multiple technical solutions worldwide for plug and socket systems in houses, it was a common goal to find a simpler solution for the electric vehicle plug and socket system. From a technical point of view, three electrical systems are available worldwide at different voltage levels: single and three-phase systems and DC. The following voltage levels are used for single-phase connectors: 110 V, 125 V, 170 V, 230 V, 240 V and 250 V. For three-phase connectors the following voltage levels are used: 190 V, 220 V, 300 V, 400 V, 420 V and 440 V. It was agreed that maximum voltages for single phase would be 250 V and for three phase would be 480 V.

The type 1 plug was proposed by Japan based on its experiences with electric vehicles. This plug covers single-phase charging of electric vehicles with a maximum current of 32 A and a maximum voltage of 250 V. Italy proposed a single phase/three phase plug with a maximum current of 32 A and a maximum voltage of 400 V.

The type 2 plug, proposed by Germany, combines AC and DC charging. The AC single- and three-phase connector can carry a maximum of 63 A at a maximum 480 V or, with the same connector, a maximum current of 70 A DC. With a connector extension the plug can be used up to 200 A DC. Parts 1 and 3 of the standard for plugs and sockets of electric vehicles was published in 2014 and Part 2 in 2011.

Table 7.13 Types of plugs.

Type 1	Single phase, max. 32 A, max 250 V (AC) Proposal: Japan
Type 2	Single/three phase, max. 63 A (AC) and 70 A (DC) max. 480 V, extendable to 200 A (DC) current Proposal: Germany
Type 3	Single/three phase, max. 32 A (AC), max. 400 V Proposal: Italy

7.3.8 Safety

The safety aspects of electric vehicles are very important for people in and around the vehicle and the equipment itself. The safety aspect is related to electrical safety, including protection against electric shock, overload conditions, fire prevention and the correct installation of low-voltage electrical installations.

Electromagnetic compatibility (EMC) is important for the correct interaction of electrical equipment in the vehicle or the charging station. Incorrect functions may cause accidents and are a danger.

There are also safety considerations related to the location of charging stations, location of parking, accurate and easy-to-understand instruction notes, labelling of the devices to be used – for example plugs and sockets. Here, recommendations to architects for parking lots, housing, office buildings, supermarkets, gas stations and any other opportunities to combine civil construction with electromobility and battery charging are needed. This infrastructure is important for the use of electric vehicles and needs long-term strategic planning.

In thinking about processes, functional safety must be ensured for all interactions around electromobility. This concerns the vehicle and the charging station itself as well as the vehicle-charging station and vehicle-vehicle communication to guarantee at any time the functional safety of the whole process.

Lightning and electrical surges occur around electric vehicles and charging stations. It is important to ensure safety under these conditions too. For most of these safety requirements standards are already available or are under preparation as shown in Table 7.14 for electric safety, Table 7.15 for electromagnetic compatibility and lightning and electrical surge protection and Table 7.16 for functional safety.

The standardization on electrical aspects of the electric vehicle and charging station is the responsibility of the IEC, its work being shared with the ISO. The main focus of the electric safety standards is on safety of individuals, protection of the equipment, handling of the charging process and communication in and around the vehicle.

IEC 61140 provides information and requirements related to protecting individuals against the danger of an electric shock by using the correct measures to install electric equipment in the vehicle and the charging station.

IEC 60479 considers the effect on human beings and livestock of an electric current passing through the body. Here maximum values of voltages and currents are defined to protect persons and livestock from harm that could come from electric vehicles or charging stations. Electric installation, according to IEC 60364, must follow requirements regarding maximum allowed voltage and currents coming from the electric vehicle, even in case of technical failures. The safety of personnel always comes first.

Table 7.14 Standards on safety of electromobility.

IEC 61140	Protection against electric shock – common aspects for installation and equipment	2001
IEC 60479	Effect of current of human beings and livestock	
	Part 1: General aspects	2005
	Part 2: Special aspects	2007
	Part 3: Effects of current passing through the body of livestock	1998
	Part 4: Effects of lightning strike	2011
	Part 5: Touch voltage threshold values physiological effects	2007
IEC 60364	Low voltage electric installation	
	Part 4-41: Protection for safety – protection against electric shock	2005
	Part 7-722: Requirements for special installations or locations – supply for electric vehicle	2015
IEC 61850	Communication networks and systems for power utility automation	2015
IEC 61851	Electric vehicle conductive charging system	
	Part 1: General requirements	2010
	Part 21: Electric vehicle for conductive connection to an AC/DC supply	2001
	Part 22: AC electric vehicle charging station	2001
	Part 23: DC electric vehicle charging station	2014
	Part 24: Digital communication between DC photovoltaic charging station and an electric vehicle for control of DC charging	2014
IEC 62196	Plugs, socket outlets, vehicle connectors and vehicle inlets - Conductive charging of electric vehicles	
	Part 1: General requirements	2014
	Part 2: Dimensional compatibility and interchangeability requirements for AC pin and contact-tube accessories	2010
	Part 3: Dimensional compatibility and interchangeability requirements for DC or AC/DC pin and contact-tube vehicle couplers	2014

Table 7.15 Standards on electromagnetic compatibility (EMC), lighting and surge protection.

IEC 61000	Electromagnetic Compatibility (EMC)	
	Part 6-2: Generic standards – immunity for industrial environments	2005
	Part 6-3: Generic standards – emission standard for residential, commercial and light-industrial requirements	2006
IEC 61851	Lightning and overvoltage protection	
	Electric vehicle charging system – overvoltages of category II	2010

Table 7.16 Standards on functional safety.

IEC 61508	Functional safety of electric/electronic/programmable electronic safety-related-systems	
	Part 1: General requirement	2010
	Part 2: Requirements for electric/electronic/programmable electronic safety-related systems	2010
	Part 3: Software requirements	2010
	Part 4: Definitions and abbreviations	2010
	Part 5: Example of methods for the determination of safety integrity levels	2010
	Part 6: Guidelines of applications	2010
	Part 7: Overview of technics and measures	2010

IEC 61850 defines the exchange of information between the communication network and system and the electric power supply by the utility. This automated process of battery charging or feedback of electric energy to the network needs to be controlled by the electric power utility and the standard must guarantee that different technical designs of different manufacturers are able to communicate with each other.

The charging process of an electric vehicle, with different methods and AC or DC current, is explained in IEC 61851 and IEC 62196 gives dimensional requirements for the plugs and sockets. This is necessary so that electric vehicles can be connected to the charging station.

Electromagnetic disturbances may come from lightning when driving through a thunderstorm, or from induced voltages when driving along a high-voltage line, or an underground cable that is not visible, or a radio frequency transmitter, or from other vehicles – for example, the ignition spark of a combustion motor car. There are many sources of surges of many different types and frequencies. The electric vehicle is designed by applying the requirements given in IEC 61000 and IEC 61851 as shown in Table 7.15.

Electrical aspects of functional safety are related to electric devices, electronic equipment and the programmable part of the electronic equipment in the electric vehicle or the charging station. This is regulated by IEC 61508 and covers safety requirements for the devices, equipment and the software. A malfunction might cause an accident and this must be avoided. Before the electric vehicle does something wrong, it should stop.

7.3.9   Ongoing Developments

Standardization of electromobility will continue in coming years. Some trends can be seen today. The close cooperation of IEC and ISO will continue to provide the full spectrum of electrical and nonelectrical standards.

The international focus of standards for electromobility is seen as a primary requirement for opening a global market for electric vehicles and charging stations. This will allow the manufacturers to serve a larger market, enabling them to meet the cost of these expensive developments. International standardization must follow technical developments swiftly. Fast standardization is essential. The stakeholders of electromobility standards need to look for clear and explicit text and requirements that make standards free from conflicts. Sometimes this is not easy as there might be conflicting industrial interests. The standardization processes of ISO and IEC offer ways to find consensus. Industrial consortia should be brought into the international standardization of ISO and IEC. Regional standardization, for example IEEE in North America or EN in Europe, must be coordinated with the standardization of ISO and IEC to avoid conflicting requirements.

Energy countries like China and India need to be integrated into the international standardization of ISO and IEC. Experts from these countries must be nominated to the related working groups to participate directly in the content of the standards and to bring this information to their national standardization organizations.

In the technical field of electrical safety four main aspects are seen, which will be subject of further standardization: the safety of the charging station, the safety of the high-voltage onboard power supply for the electric vehicle, the wiring of the electric vehicle and the electrical, chemical and mechanical safety of the battery, even in an accident. With the increasing use of digital communication, automated driving and many other services on the electric vehicle the electromagnetic compatibility of all systems and devices will also be a focus of standardization.

More standardization is needed on external interfaces to the vehicle, the power supply networks, the charging station, the electricity provider and trader, the provider of the power electric infrastructure, the user of the electric vehicle and any service related to electromobility.

One of the most urgent standardization activities is the coordination of the communication requirements of electric vehicles and charging stations with Smart Grid requirements. Both systems need to be able to communicate and cooperate.

The aspect of load management for charging and the back feed of electric power to the network needs standardization for timing, load control and price control. This is a complete new business segment, which can only be developed in the case of standardization.

Then there are aspects of standardization related to the situation of a power supply interruption and the rebooting of the system, which needs standards to coordinate all the different players. Communication standards for diagnostics combined with services are also coming and, with these standards, new business models.

Finally, the external connections to AC or DC charging stations must be continuously adapting to technical developments. The standards regarding the functional safety of electric vehicles and charging stations also need to be updated continuously to take into account technical developments, including standards on information technology, safety and data privacy.

For electric vehicle performance and energy consumption, standards are required on measurement methods and ways to compare different manufacturer designs. This includes the electric vehicle, the battery system and the charging infrastructure.

The situation of the electric vehicle and the battery system after an accident involving total or partial damage needs standardization. In the near future standards on lifecycle evaluations and battery testing will be available. The open and accessible handling of a large number of electric vehicles when the batteries of these vehicles are connected by control data links over the Internet energy storage or to deliver electric energy to the power supply network will need a new set of standards. This will create new ways of doing business in a new technological environment.

The so-called Ultracaps, as a fast, low-loss, highly flexible, high-power electric storage system will then be the next to be standardized.

Electromobility not only relates to the electric vehicle and the charging station. It also influences our living and society. It will impact the way we will build our homes or large apartment buildings, which will need to provide charging stations for the electric vehicles and include services for handling the metering and billing by the facility management or any service provider. Parking lots in cities in the future will provide charging services while the electric vehicle user enjoys shopping. Electromobility will give architects and city planners new options in offering integrated services around electromobility. Power-supply networks will change their structure with more intelligent control and measurement options. Electric installations will change to meet the needs of electric vehicles with fast electric charging stations.

The average charging requirement for electric vehicles depends on weight and size and is between 10 and 20 kWh. This current can be provided as AC or DC and will require charging times of about 5–10 h when only 2 kW can be charged from a single-phase AC connection, or about half an hour to an hour with a three-phase AC connection with up to 40 kW of charging power. With a high-power DC charging connection of up to 170 kW, the charging time goes down to some minutes. This wide spectrum shows the impact of the infrastructure for charging the battery of an electric vehicle. Not many people will buy an electric vehicle if the charging time is in the range of 5–10 hours – it is too long for practical use. This picture is very different if the time is down to 10–20 minutes. Here, beside standardization, financial investment into the infrastructure needs to be planned.

7.4 Conclusion

The two case studies chosen, on Smart Grid and electromobility, show how international standardization is forming solutions influenced by the market and trends in technical developments. Both technical fields started under very different conditions. The Smart Grid idea was born from digital thinking and the idea that, through data handling, measuring and sensoring, digital control, digital analysis and digital services using the Internet, the classical power supply network could be made much smarter.

But what is smart? The worldwide search for smart power supply network ideas produced many suggestions. Control of washing machines by the power utility with a smart meter at times of low electricity prices (usually at night). Control of network stability when using fluctuating power-generation sources, such as wind and sun, with thousands of smart algorithms for network control and simulation (every IEEE conference is full with papers in this field). Control of house lights and heating using the smart phone from any location on the planet. Control of electricity bills by smart software. Control of the electric power generated by the rooftop photovoltaic panels with smart control to sell the electricity to the power utility or to use by the building's occupant. Smart Grid application of an independent micro grid where are small electrical islands of electric power generation (photovoltaic, wind, bio mass, running water hydro plant, hydro storage and diesel generator) to be kept in balance with the electricity user in this micro grid (housing, apartments, industry, agriculture, hospitals), all interconnected and operated in an optimized way to use the lowest cost energy first.

And it works in many places throughout the world. This wide field of innovative technology and the great need for standardization worldwide started in many locations, with different focuses and at different speeds. But after some years, through the coordinating forces of standardization, different organizations learned from each other. The participants in standardization activities for Smart Grid and e-mobility were able to avoid unsolvable conflicts and found solutions for the standards published.

The IEC was the driving organization in this coordination work, pushed by a worldwide community and regional organizations like the IEEE in North America, CENELEC in Europe and the SAC in China. Most regional activities were brought up to IEC international level and then found its way into international standardization. This helped to create a harmonized technical level IEC 61850, which today is the global accepted standard for digital communication in automated electrical power networks.

Not every technical solution has found its way to international level – specific regional requirements need to be addressed – but the wide basis of IEC standards cover most cases.

Smart Grid development is not at an end. There are still rapid and basic technical innovations in this field, so the need for standards is still high. The success of international standardization depends very much on the will of the global family to be open to work with the transparent process of standardization. If single companies become too strong and try to set standards with their products then no consensus-based standards can be produced. We have seen this with Microsoft's dominance in the operation software of personal computers. Future standards may come from Apple, Google and Facebook in some technical areas. The world-wide acceptance of IEC, ISO or ITU as technical standardization organization is expected to dominate the technical world.

The second case study was very different in terms of the standardization processes. It was clear from the very beginning what an electric vehicle is – a car with a battery and an electric motor. The challenge was and is to coordinate the standardization between two very different industrial sectors – the automotive industry, with large global companies, and the electric industry, also with global activities. The strategic plans of both industries have been very different, related to their type of business. The automotive industry sees a replacement business when electric vehicles will replace vehicles with combustion motors. In the beginning the electric vehicles might be used alongside the classic car. The electric industry is entering a new business with new products and services. The complexity of standardization around electric vehicles and charging stations is very high, requiring electrical, mechanical and chemical fields of expertise. Close coordination between the ISO, IEC and ITU, right from the beginning, formed a good basis for international standardization.

Regional contributions to international consensus standards (e.g. plugs and sockets from Italy, Japan and Germany) and the will of all participants to contribute to the international standardization have helped to form a wide spectrum of standards. It can be expected that this international approach will continue and that the innovative technical field of electromobility can create the necessary international standards.

The goals and the role of the automobile industry and how fast the electric vehicles sector will develop are important factors. The large investment of Tesla into the production of lithium-ion batteries will also affect the speed of development of standards around electromobility.

More case studies are possible that could show the strong impact of standards on technical developments and business success. Wireless communication based on IEEE 802 is one example of how a specific technology is spreading around the globe with the required standards. The close link between the standardization process and technical development made this standard a success story and this helped businesses using wireless communication.

Ultra-high voltages (UHV) are defined as voltages of 1 000 000 V and above for the purpose of high-power transmission over long distances. This technology requires large and very expensive equipment like transformers, switchgear and transmission lines. To design such equipment without standards would lead to very expensive project orientated design requirements. Therefore, an international standard based on only a few projects in China was the basis for formulating the requirement for the design and testing of UHV equipment.

These and other case studies around cybersecurity, ambient assisted living and smart cities are planned for the second edition.

7.5   Publicity in Japan

Standards in Japan are being promoted by public media on TV and radio. For example, the largest TV station in Japan, NTK, used prime TV time – 8.00 p.m. on a Saturday evening – to promote the need for international standardization in a 45 minutes broadcast.

Ultra-high voltage equipment is used for electric power transmission with voltages higher than a million volts – a really high voltage rating! Such equipment has been developed in different countries to transmit high power ratings over long distances. Some installations have been developed for testing in Italy, the United States and Russia and some installations went into service but never used the maximum voltage in operation. Japan also developed such UHV technology but never operated it at one million volts. The latest development in UHV was started in China including the erection of a 600 km-long transmission line. China then started its standardization initiative.

In Japan this was recognized as a threat to Japanese industry and a disadvantage to export this technology to the world market. This was the situation when the idea to the TV show was born. The TV show or report had the structure of a Samurai movie. Three brave and engaged Japanese experts from manufacturers and users in Japan travelled around the world and tried to convince the standardization community to follow the Japanese test voltage values and process in the standard document.

The TV broadcast started with an overview of standardization in Japan and where specific Japanese technical solutions are not used in the world market, which follows international standards such as ISO or IEC. One example is the efficiency rating for refrigerators with A, A+, A++, B, C, which is not available in Japan and so it is almost impossible for Japanese refrigerators to be sold outside Japan.

Another example is the mobile phone business, where Japanese technology does not conform to international standards. Many more examples could be mentioned where Japan is following Japanese standards for the home market and international standards for the global market. For Japanese industry this means that different equipment is manufactured in different factories for the domestic and world market, that there are different maintenance concepts and different rules for operation and end-of-lifetime use. This protects the home market because most manufacturers follow the requirements of international standards and avoid special solutions for a small domestic market.

This situation was explained to the public in Japan, together with the way in which a chance to be successful and profitable on the world market was being missed as a result.

The main part of the broadcast showed the standards development process within the international experts committee and the discussions from the first draft to the final publication. The three Japanese experts followed the meeting together with the NTK TV team to the locations of meetings first in Sweden, then in India, then in Brazil, then in Japan, then in the United States and finally in Germany. At each step it followed the progress of a very technical document with power frequency test voltage ratings, lighting impulse voltage ratings and many other technical values of importance for Japan.

In this particular case the danger for Japan would be if the higher values promoted by China had been taken into the standard. This would have closed the door to the world market. Japanese industry would have had to redesign the UHV equipment at a very high cost.

The TV broadcast promoted the three experts from Japan as brave fighter for the Japanese technique, closely following the development of the standard's text. Japan played a very active role. Other experts in the TV broadcast had a role in opposing the Japanese proposals. Germany, the United States, Sweden, Switzerland and others in the broadcast were critical of the Japanese goals. The broadcast brought some entertaining aspects to the information covered.

As there is no English translation of the broadcast, my Japanese expert colleague on the IEC working group came to my home to explain its content to me. He had to bring a Japanese video player because the CD was designed to conform to Japanese standards and cannot be played on an international video player – another good reason why this public information on the importance of international standardization is so important.

8

Conformity and Certification

8.1 General

What is a Conformity Assessment?

To conform with a standard, its users must follow its requirements and test procedures. Standard requirements are usually clear but fulfilling the test requirements in the standard depends on some conditions.

First there is the quality of the test facility. Does the test generator produce the required voltage, current, heat, cold, mechanical force and so forth in a way that fulfils the requirements for the test setup?

Second there are the qualifications of the test laboratory staff. Do the test engineers have the right qualification and experience in setting up tests that fulfil the standard's criteria?

And third: does the test laboratory use the correct measurement equipment, with the required sensitivity and accuracy for the measurements?

All of these criteria are part of the conformity assessment and certification.

8.2 Processes

Standards contain requirements and test procedures to prove that the requirements have been fulfilled by the tested device or system. Conformity assessment gives rules to those who carry out the test procedures. Then the test results are documented in certificates.

Conformity assessment plays an important role for standardization and its practical application. Test laboratories need to follow the same rules at the same test quality level to avoid a situation in which the same device following the same standard would pass in one test laboratory and fail in another. The conformity assessment process has therefore been established as shown in Figure 8.1.

Practical Guide to International Standardization for Electrical Engineers: Impact on Smart Grid and e-Mobility Markets, First Edition. Hermann J. Koch.
© 2016 John Wiley & Sons, Ltd. Published 2016 by John Wiley & Sons, Ltd.

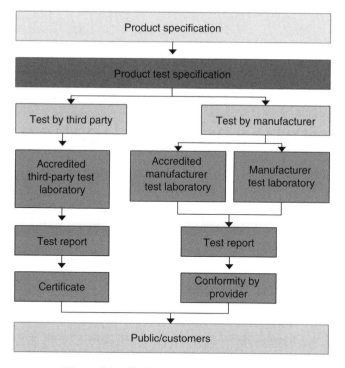

Figure 8.1   Conformity assessment process.

The principle of the conformity assessment process is that the manufacturer of devices or systems indicates to the public or customers that his products are fulfilling the requirements given in the related standards.

The first decision of the product manufacturer is whether the tests will be made by the manufacturer or by a third-party test laboratory. Test laboratories must be accredited by an independent certifier to prove that the test conditions, quality of test equipment and knowledge of test engineers are able to provide correct test results. The test engineers also need to be independent from the product manufacturer in their position in the test laboratory.

In any case, it is also possible in the conformity assessment to use the internal test laboratory of the manufacturer to provide test protocols for the device or system even when the test laboratory of the manufacturer is not certificated. The question in this case is whether the customer will accept these test protocols.

After the tests have been passed, the certificate will prove the conformity of the product with the related standards.

The manufacturer then uses the conformity certificates to indicate to the public and the customer that the tested devices and systems conform with the standards (www.iecq.org, accessed 7 February 2016).

The conformity evaluation scheme has a modular structure and offers different modules according to specific requirements as shown in Figure 8.2.

In Module A the certificate covers the complete internal production process of the factory assembly plant. In Module B the specimen test is certificated in general while Module C gives

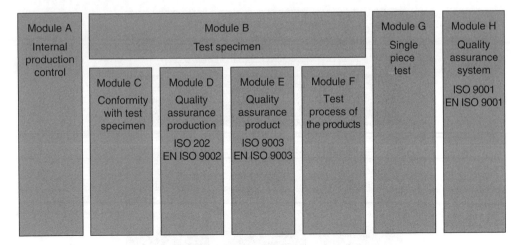

Figure 8.2 Modular structure of conformity evaluation.

certification for one specimen tested. Module D tests the production process (ISO 9002), Module E the product itself (ISO 9003) and Module F the test process of the product. In Module G, processes for single-piece tests are certificated and in Module H the overall quality assurance system (ISO 9001) is certificated.

8.3 IEC Process

Conformity assessment plays an important role for the IEC. The Conformity Assessment Board (CAB) is a separate organization in the IEC. The members of IEC CAB are delegated by the IEC council. The CAB is a decision-making body and reports to the council. The organization of IEC CAB is shown in Figure 8.3.

The Conformity Assessment Board (CAB) has three subdivisions – on electronic components, electrical equipment and explosive atmospheres.

IEC Q

The CAB on electronic components (IEC Q) is a worldwide approval and certification system covering the supply of electronic components and associated materials and assemblies (including modules) and processes. It uses quality-assessment specifications based on IEC standards and publishes brochures on the following topics:

- *IEC Q assures quality in the electronic industry;*
- *IEC Q Increased car safety and reliability;*
- *IEC Q Protect your investment, reduce your liability;*
- *IEC Q approved process schemes;*
- *IEC Q approved component products, related materials and assemblies scheme;*
- *IEC Q demonstrates compliance with new and stricter hazardous substances regulation;*
- *IEC Q scheme for LED lighting;*
- *IEC Q hazardous substance process management (HSPM).*

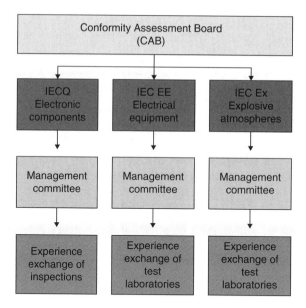

Figure 8.3   IEC Conformity Assessment Board (CAB).

The work of IEC Q is to accept the quality of electronic components without any additional tests. This means that electronic components that have been tested once do not need to be tested again when used in electronic products in any subsequent manufacturing process.

The IEC Conformity Assessment Board operates seven subdivisions on the topics discussed below (www.iecq.org, accessed 7 February 2016).

IEC Q AP Scheme

The IEC Approved Process Scheme (IEC Q AP) covers any processes related to electronic components, assemblies and services. For example, this may cover, among other areas, product engineering, printed wiring board manufacturing, electronic component manufacturing, printed circuit board assembly, electrostatic discharge (ESD) controls or even supply-chain management.

The approved process scheme permits organizations to certify their specialized services or processes for manufacturers to certify or qualify them.

Organizations that qualify for the IEC Q AP scheme and hold a certificate demonstrate to the international market place that their organization and facilities comply with the requirements of ISO 9001 (www.iecq.org/about/ap-scheme, accessed 7 February 2016).

IEC Q AC Scheme

The scope of the IEC Approved Component Products, Related Materials and Assemblies Schemes (IEC Q AC Scheme) is to cover requirements formed in a standard or manufacturer specification in the following technical fields, among others: silicon, wafer slaps, integrated and discrete electronic components, connectors, printed wiring board products/components/ materials that assist in the construction, installation and use of electronic components.

Organizations that are holding an IEC AC Scheme certificate indicate to the international market place that they follow the IEC Q AC Scheme and the relevant technical standards and specifications for their activity (www.iecq.org/about/ac-scheme, accessed 7 February 2016).

IEC Q AQP Scheme

The scope of the IEC Q Automotive Qualification Program (IEC Q AQP) gives the automotive industry a standardized way to test the components to ensure their reliability.

The IEC Q AQP provides the automotive industry with an efficient way to ensure that the components they are buying meet expected quality, safety and reliability requirements. The IEC Q AQP ensures that the sampling, test processes, test results and production controls are monitored and verified by an impartial third-party certification body.

Organizations that hold IEC Q AQP certificates demonstrate to the international market place, through testing and other verification criteria, that they and their facilities comply with the requirements of the relevant standards and specifications (www.iecq.org/about/app, accessed 7 February 2016).

IEC Q Avionics Scheme

The IEC Q Avionics Parts and Assembly Management requirements are designed to evaluate commercial, military and aerospace avionics equipment manufacturers and related organizations. The management plans are used to develop, document and implement plan owners' processes for managing the selection and use of electronic components in avionics equipment.

Organizations that hold IEC Q avionics certifications demonstrate to the international market place that their organization and facilities comply with the requirements of IEC TS 62239-1 or GEIA 4899 (www.iecq.org/about/avionics, accessed 7 February 2016).

IEC Q CAP Scheme

The scope of IEC Q Counterfeit Avoidance Programme (IEC Q CAP) allows all organizations, including equipment manufacturers or subcontractors that the organizations use, to develop and implement the system required to obtain an IEC Q CAP certificate of conformity (www.iecq.org/about/cap, accessed 7 February 2016). This certificate demonstrates internationally that the organization has processes to managing counterfeit avoidance in the selection and use of components according to IEC Q CAP technical and quality management system requirements. This is ensured through independent conformity assessment and ongoing surveillance by an IEC Q-certified body.

Organizations that hold IEC Q CAP certification demonstrate to the international market place that their organization and facilities comply with the requirements of the IEC Q System for Counterfeit Avoidance (www.iecq.org/about/cap, accessed 7 February 2016).

IEC Q HSPM Scheme

The scope of the IEC Q HSPM Hazardous Substance Process Management (IEC Q HSPM) requirements are designed to evaluate equipment manufacturers' and related organizations' processes for compliance with QC 080000 IECQ HSPM (IEC Q HSPM) in addition to the compliance of their processes with ISO 9001 QMS.

Organizations that hold an IEC Q HSPM certificate demonstrate to the international market place that the organization has developed, documented and implemented processes for managing the production, selection and use of electronic components, assemblies, processes and related materials in accordance with customer, local, national and international health-and-safety requirements for their scope of activity (www.iecq.org/about/hspm-scheme, accessed 7 February 2016).

IEC Q ITL Scheme

IEC Q Independent Testing Laboratory (IEC Q ITL) approval is available to independent test laboratories intending to carry out tests in support of IEC Q activities within the IEC Q system. The approval covers the type of tests to be carried out, the component ranges to be tested and the facilities available and exceeds the relevant requirements of ISO/IEC 17025.

Independent testing laboratories that hold IEC Q ITL certificates demonstrate to the international market place that they and their facilities comply with IEC Q requirements for the competence of staff and the adequacy of testing facilities and that they perform their functions under the IEC Q System (www.iecq.org/about/itl-scheme, accessed 7 February 2016).

IEC EE

The second column in Figure 8.3 shows the IEC of a conformity assessment scheme for electrotechnical equipment and components (IEC EE) (www.iecq.org/about/itl-scheme, accessed 7 February 2016) and is a subdivision of the IEC CAB. The IEC EE has published a certification body scheme (CB scheme) to improve international trade in electrical equipment for homes, offices, workshops and healthcare facilities. The CB schemes benefit customers, industries, authorities, and so forth, to provide confidence in manufacturers and services provided by various national certification bodies (NCBs) in applying rules.

The CB scheme is based on the principle of mutual recognition by its members of test results for obtaining certification or approval at national levels.

The CBS is also based on the use of certification body test certificates, which provide evidence that the representative specimen of the product has successfully passed the test and showed compliance with the requirements of the relevant IEC standards. The results are documented in the certification body test report.

The national certification body (NCB), intending to operate in the CBS, needs to be recognized as a national certification body by IEC EE.

The IEC Q EE has today 56 member countries through their national member body committees.

Behind the national members a wide range of organizations can be found as shown in Table 8.1.

Each test is only done once in an IEC EE certified test laboratory and will then be accepted by any participating country. On the basis of a certificate from a national certification body in one location (certification body A) the manufacturer can get the certificate and certification sign of another (certification body B). Certification body B can ask for a certification specimen to repeat the test. Either certification body A or certification body B can provide a certificate, if the certificate is based on the CBS. Certification body B may monitor the manufacturer's facilities if the local regulations cover this aspect.

Table 8.1 Organizations behind the national members of IEC Q EE (some examples, status 2015).

Nation	Members
Germany	VDE Prüf- und Zertifizierungsinstitut
	TÜV Rheinland LGH Products
	TÜV Süd Product Service
	TÜV Nord CERT
	SLG Prüf- und Zertifizierungs GmbH
	Eurofins Product Services
	TÜV Inter Cert
	Büro Veritas
	Hansa Control
	Intertek
France	LCIE – Bureau Veritas
	LNE – Colaborateurs Laboratoire national de mètrologie et d'essais
	Apave Certification
India	BIS – Bureau of Indian Standards
	STQC – Ministry of Communications and Information Technology
Italy	IMO – Instituto Italiano del Marchio di Qualita
	ICIM – Marchino di Certificone
Japan	JET – Japan Electrical Safety and Environment Technology Laboratories
	JQA – Japan Quality Assurance Organization
	TÜV Rheinland Japan
	UL Japan
China	CQC – China Quality Certification Centre
Korea	KTL – Korea Testing Laboratory
	KTC – Korea Testing Certification
	KTR – Korea Testing and Research Institute
	KERI – Korea Electrotechnology Research Institute
	NERC – New and Renewable Energy Centre
Netherlands	DEKRA – Certification B.V.
	Kiwa Netherlands B.V.
Russia	GOST Re – Federal Agency of Technical regulatory and metrology
United Kingdom	Intertek Testing and Certification Ltd.
	BSI – British Standards Institute
USA	UL – Underwriters Laboratory
	MetLaboratories – Electric Test and Certification Laboratory
	ITS – Intertek Testing Services
	TÜV – Rheinland North America

IEC Ex System

The third column of Figure 8.3 shows the IEC System for Certification of Standards Relating to Equipment for use in Explosive Atmospheres (IEC EX System).

The objective of the IEC Ex system is to facilitate international trade in equipment and services for use in explosive atmospheres, while maintaining the required level of safety. The goals are to reduce testing and certification costs for the manufacturer, reduce time to market, to promote international confidence in the product assessment process and one international database listing and maintain international confidence in equipment and services covered by IEC Ex Certificates.

Typical areas for IEC Ex systems are found in automotive refilling stations, oil refineries, chemical processing plants, printing, the paper and textile industry, aircraft refilling and hangars, the surface-coating industry, underground coal mines, sewerage treatment plants, gas pipelines, grain handling and storage, woodworking areas, sugar refineries and metal surface grinding.

IEC Ex publishes rules for administering the IEC Ex System. There is the standards series 60079 on explosive atmospheres, series 61241 on electric apparatus for the use in presence of combustible dust, series 62013 for cap light for use in mines susceptible to fire damp and series 62089 for electrical apparatus for explosive gas atmospheres.

The certificates of the member countries of IEC Ex are accepted for the certification sign. Within the European Economic Area the Explosive Protection Directive is applied. A process for direct acceptance of IEC Ex Certificates is under preparation.

8.4   European Process

European Conformity is using the CE sign to indicate that the product or service or system is in accordance with all related EN standards and EU Directives.

The CE sign shows that all protection goals of the standards and directives have been fulfilled. The sign indicates that the manufacturer or the distributor is in accordance with any related EN standard or EN Directive. When the product bears the CE sign it has free market access to the European Union without any additional tests or checks made by the European Union. The responsibility for the correct use of the CE sign is the manufacturer's or distributor's.

The CE sign is meant to be used by the national market access control, which is usually linked to the Ministry of Commerce in the EU member countries. The sign is not meant to inform the consumer. The market access control will check if the CE sign is indicated on the product and in cases of doubt may test the product according to the related standards. Exclusion from the market will follow if the tests are not passed.

The responsibility for the correctness of the CE sign is with the manufacturer and/or distributor or anybody working on their behalf. The CE sign is not a quality certification or a safety indication.

ENEC

The ENEC is a high-quality European safety mark for electric products that demonstrate compliance with European standards (EN). European certification bodies in the electrical sector have opened the European ENEC mark to all electrical product sectors. Testing takes place in independent ENEC-approved testing laboratories worldwide and in approved manufacturers' testing laboratories. ENEC means safety in Europe and all signatories to the ENEC mark scheme actively support its commitment to higher safety levels (www.enec.com, accessed 7 February 2016).

HAR Agreement

The HAR Agreement is for commonly agreed markings of cables and cords complying with harmonized specifications and European standards.

It was established in 1974 and covers the low-voltage field (<1000 V) for cables and cords.

Member countries are Austria, Belgium, Czech Republic, Finland, France, Germany, Hungary, Italy, Netherlands, Norway, Poland, Portugal, Spain, Sweden, Switzerland, Turkey, and the United Kingdom.

The HAR sign gives confidence in choosing the right cable for specified applications using EN standards and harmonized CENELEC documents (HD). It is based on more than 40 years of experience with the HAR system. The HAR system is used by local (national) electric installation inspection and regulation organizations, national regulations, codes and standards in the EU, manufacturers' data sheets, CENELEC guides to cable use (EN 50565-1 and EN 50565-2) and EN standards for cable products (www.iec.ch; https://standards.ieee.org, both accessed 7 February 2016).

LOVAG

The Low Voltage Agreement Group (LOVAG) was founded in 1991 and has been recognized as a certification body by Italy, Germany, France, Belgium, Spain and Sweden. The objective of LOVAG is the harmonization of testing and certification for low-voltage industrial, commercial and similar electric appliances. The certificates of any national certifying institute will be accepted by the other national certifying institutes. Only one certificate is required for the entire EU market.

LOVAG in the European Compliance Scheme provides global access to the largest world-wide market for low voltage industrial products by using international standards for product certification (www.iec.ch; http://trade.gov/mdcp/, both accessed 7 February 2016).

Public Legal Market Access in EU

In the European Union products and services can only be brought to market if:

- The products or services are in accordance with EU Directives. For safety and health requirements it is assumed that the product is used under normal application conditions and with reasonably foreseeable misuse of the product.
- If there is no EU Directive with a sign requirement. In this case the products or services can only be brought to market if the product is used under normal conditions and with reasonably foreseeable misuse and no health and safety risks are expected.

These requirements for market access are regulated by the member countries of the EU with national laws – in Germany, for example, by § 4 of the Produktsicherheitsgesetz (www.iec.ch, accessed 7 February 2016; http://www.nist.gov/, accessed 3 March 2016).

References

[1] American National Standards Institute (2016) *ANSI Essential Requirements: Due process requirements for American National Standards*, http://publicaa.ansi.org/sites/apdl/Documents/Standards%20Activities/American%20National%20Standards/Procedures,%20Guides,%20and%20Forms/2016_ANSI_Essential_Requirements.pdf (accessed 2 March 2016).

[2] Blind, K., Jungmittag, A. and Mangelsdorf, A. (2010) *Gesamtwirtschaftlicher Nutzen der Normung*, DIN Deutsches Institut für Normung e. V., Berlin, http://www.din.de/blob/79542/946e70a818ebdaacce9705652a052b25/gesamtwirtschaftlicher-nutzen-der-normung-data.pdf (accessed 2 March 2016).

[3] Deutsche Kommission Elektrotechnik Elektronik Informationstechnik in DIN und VDE (2014) *Annual Report 2014*, DKE, Frankfurt-am-Main, https://www.vde.com/en/dke/annualreport/Documents/DKE_JB2014_EN_PDF.PDF (accessed 2 March 2016).

[4] IEEE 2030-2011 (2011) *IEEE guide for Smart Grid interoperability of energy technology and information technology operation with the electric power system (EPS), end-use applications, and loads*, Institute of Electrical and Electronics Engineers, New York, NY.

Practical Guide to International Standardization for Electrical Engineers: Impact on Smart Grid and e-Mobility Markets, First Edition. Hermann J. Koch.
© 2016 John Wiley & Sons, Ltd. Published 2016 by John Wiley & Sons, Ltd.

Index

Practical Guide to International Standardization for Electrical Engineers: Impact on Smart Grid and e-Mobility Markets, First Edition. Hermann J. Koch.
© 2016 John Wiley & Sons, Ltd. Published 2016 by John Wiley & Sons, Ltd.